AF510766

PRINCIPLES OF OCEANOGRAPHY

Edited by
Judith Rosales

Delve Publishing

Principles of Oceanography

Edited by Judith Rosales

© 2017

Edition 2018

Delve Publishing

Published by

Delve Publishing LLC

708 3rd Avenue, 6th Floor

New York

NY 10801

United States of America

www.delvepublishing.com

ISBN: 978-1-68095-736-5 (Hardback)

Printed in United States of America

Library of Congress Control Number: 2016920416

This book contains information obtained from highly regarded resources. Some content has been published with permission under Creative Commons License or similar. A Wide variety of references are listed. Reasonable efforts have been made to publish reliable data and views articulated in the chapters are those of the individual contributors, and not necessarily those of the editors or publishers. Editors or publishers are not responsible for the accuracy of the information in the published chapters or consequences of their use. The publisher believes no responsibility for any damage or grievance to the persons or property arising out of the use of any materials, instructions, methods or thoughts in the book. The editors and the publisher have attempted to trace the copyright holders of all material reproduced in this publication and apologize to copyright holders if permission has not been obtained. If any copyright holder has not been acknowledged, please write to us so we may rectify.

Notice: Registered trademark of products or corporate names are used only for explanation and identification without intent of infringement. Although care has been taken to check accuracy of formulas and procedures, the detailed methods should be tested further on a small scale before being adopted commercially.

Delve Publishing publishes wide variety of books and eBooks. For more information about Delve Publishing and its products, visit our website at www.delvepublishing.com

About the Editor

Judith Rosales

Prof. Judith is an Ecologist and Environmental Specialist with more than 30 years of academic and consulting experience. She holds a PhD in geography from the University of Birmingham UK and wrote the PhD and Master Academic Programs of Environmental Sciences for the Universidad de Guayana, Venezuela.

Table of Contents

3 OCEAN ZONES

4 OCEANIC LANDFORMS

5 PELAGIC SEDIMENT

6 PHYSICAL OCEANOGRAPHY

7 WIND WAVE

8 COASTAL LANDFORMS AND PROCESSES

9 MARINE ECOSYSTEM AND CONSERVATION

List of Abbreviations

DOW	Deep ocean water
DSM	Deep Sea Mining
ECMWF	European Centre for Medium-Range Weather Forecasts
GCMs	Global Circulation Models
GODAS	Global Ocean Data Assimilation System
GPP	Gross primary production
HROV	Hybrid-ROV
IUCN	International Union for Conservation of Nature
MPAs	Marine protected areas ,
MTSG	Marine Turtle Specialist Group
NADW	North Atlantic Deep Water
NOAA	National Oceanic and Atmospheric Administration
NODC	National Oceanographic Data Center
PSAT or PAT	Pop-up satellite archival tag
RFID	Radio-frequency identification
ROV	Remote-operated vehicle
TEDs	Turtle excluder devices
WFD	Water Framework Directive
WRI	World Resources Institute

Preface

Oceanography is the branch of science which deals with the physical and biological properties of the ocean and the various phenomena related to it. It covers a wide range of topics, including ecosystem dynamics; ocean currents, waves, and geophysical fluid dynamics; plate tectonics and the geology of the sea floor; and fluxes of various chemical substances and physical properties within the ocean and across its boundaries." Oceanography studies the naturally happening, large-scale drifts on the Earth, the mobility of fluids in the ocean known as geophysical fluid dynamics; plate tectonics or the movements of the large sections that form the ocean floor and the processes of land formation or geology of the sea bed. The ocean experiences numerous variabilities of various chemical substances present in its water-body. The physical properties within the ocean and beyond its boundaries also affect the ocean. These changes are the subject matter of oceanography. Oceanography is an amalgamation of diverse topics and oceanographers reflect multiple disciplines like biology, chemistry, geology and physics that interrelate to expand knowledge of the world oceans and the processes within. The areas which Oceanography brings under its focus is the deep sea and the shallow coastal regions. The oceans of the world are not isolated from the rest of the physical features of the world. They are influenced by and influence the atmosphere, the continents and the frozen water part of the Earth system called the cryosphere. Together they form a harmonious whole, absorbing the sun's rays for energy. It also receives energy from tides caused by the pull of the moon, the sun and planets. The Earth's interior also radiates heat. Oceanography is inclusive of the disciplines that analyze the workings of marine life, ecosystems and ocean circulation. Oceanography is a very important discipline for understanding the oceans, which are essential for all life on Earth and cover the majority of the Earth's surface.

Content Coverage

Chapters One and *Two* present an overview of origin of oceans and seabed structure, correspondingly. The oceans of Earth are unique in our Solar System. Seabed is an experimental industrial field which involves extracting submerged minerals and deposits from the sea floor. Most of the oceans have a common structure, created by common physical phenomena, mainly from tectonic movement, and sediment from various sources

Chapters Three and *Four* present a coverage of ocean zones and coastal landforms, consistently. There are four major oceanic zones where plants and animals live in the ocean. The four major zones are intertidal zone, neritic zone, open ocean zone and benthic zone. These zones contain the largest ecosystem on Earth.

Chapters Five and *Six* are aimed to focus on pelagic sediments and physical oceanography, separately. Pelagic sediment is composed of calcareous and siliceous shells from marine organisms, fine-grained clay material, volcanic ash and trace amounts of meteor debris, however physical oceanography covers study of physical conditions and physical processes within the ocean, especially the motions and physical properties of ocean waters

Chapters Seven and *Eight* are intended to ocean surface waves and coastal landforms processes, separately. Wind waves or ocean surface waves have a certain amount of randomness: subsequent waves differ in height, duration, and shape with limited predictability. Waves erode coastline as they break on shore releasing their energy; the larger the wave the more energy it releases and the more sediment it moves.

Chapter Nine is marine resources conservation focuses on limiting human-caused damage to marine ecosystems, restoring damaged marine ecosystems, and preserving vulnerable marine species.

CHAPTER 1

Origin of Oceans on Earth

Chapter Contents

The oceans of Earth are unique in our Solar System. No other planet in our Solar System has liquid water (although recent finds on Mars indicate that Mars may have had some liquid water in the recent past). Life on Earth originated in the seas, and the oceans continue to be home to an incredibly diverse web of life.

INTRODUCTION

An ocean is a body of saline water that composes much of a planet's hydrosphere. On Earth, an ocean is one of the major conventional divisions of the World Ocean, which covers almost 71% of its surface. These are, in descending order by area, the Pacific, Atlantic, Indian, Southern (Antarctic), and Arctic Oceans. The word *sea* is often used interchangeably with "ocean" in American English but, strictly speaking, a sea is a body of saline water (generally a division of the world ocean) partly or fully enclosed by land.

Saline water covers approximately 72% of the planet's surface ($\sim 3.6 \times 10^8$ km^2) and is customarily divided into several principal oceans and smaller seas, with the ocean covering approximately 71% of Earth's surface and 90% of the Earth's biosphere. The ocean contains 97% of Earth's water, and oceanographers have stated that less than 5% of the World Ocean has been explored. The total volume is approximately 1.35 billion cubic kilometers (320 million cu mi) with an average depth of nearly 3,700 meters (12,100 ft).

As it is the principal component of Earth's hydrosphere, the world ocean is integral to all known life, forms part of the carbon cycle, and influences climate and weather patterns. It is the habitat of 230,000 known species, although much of the oceans depths remain unexplored, and over two million marine species are estimated to exist.The origin of Earth's oceans remains unknown; oceans are thought to have formed in the Hadean period and may have been the impetus for the emergence of life.

Extraterrestrial oceans may be composed of water or other elements and compounds. The only confirmed large stable bodies of extraterrestrial surface liquids are the lakes of Titan, although there is evidence for the existence of oceans elsewhere in the Solar System. Early in their geologic histories, Mars and Venus are theorized to have had large water oceans. The Mars ocean hypothesis suggests that nearly a third of the surface of Mars was once covered by water, and a runaway greenhouse effect may have boiled away the global ocean of Venus. Compounds such as salts and ammonia dissolved in water lower its freezing point, so that water might exist in large quantities in extraterrestrial environments as brine or convecting ice. Unconfirmed oceans are speculated beneath the surface of many dwarf planets and natural satellites; notably, the ocean of Europa is estimated to have over twice the water volume of Earth. The Solar System's giant planets are also thought to have liquid atmospheric layers of yet to be confirmed compositions. Oceans may also exist on exoplanets and exomoons, including surface oceans of liquid water within a circumstellar habitable zone. Ocean planets are a hypothetical type of planet with a surface completely covered with liquid.

Surface of the Atlantic Ocean meeting Earth's planetary boundary layer and troposphere, a range view which varies depending on the assumed surface elevation.

ETYMOLOGY

The word " ocean " comes from the figure in classical antiquity, Oceanus , the elder of the Titans in classical Greek mythology, believed by the ancient Greeks and Romans to be the divine personification of the sea, an enormous river encircling the world.

Earth's Global Ocean

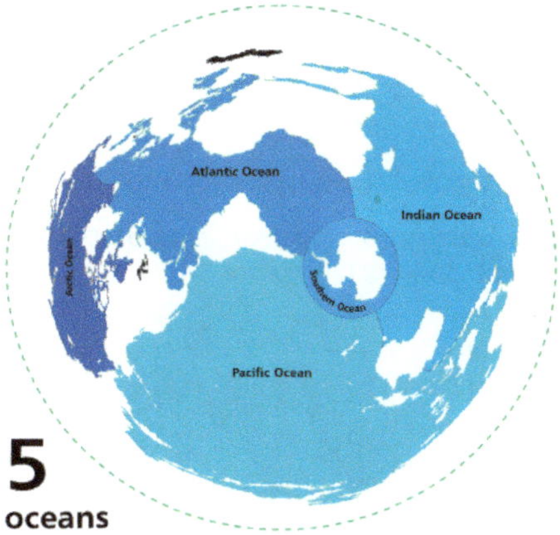

Various ways to divide the World Ocean

Oceanic Divisions

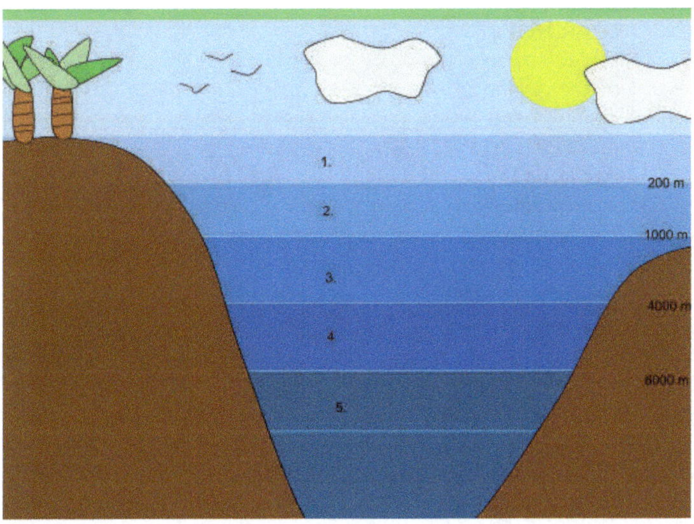

1. Epipelagic zone: surface - 200 meters deep 2. Mesopelagic zone: 200 - 1000m 3. Bathypelagic zone: 1000m - 4000m 4. Abyssopelagic zone: 4000m - 6000m 5. Hadal zone (the trenches): 6000 m to the bottom of the ocean

Though generally described as several separate oceans, these waters comprise one global, interconnected body of salt water sometimes referred to as the World Ocean or global ocean. This concept of a continuous body of water with relatively free interchange among its parts is of fundamental importance to oceanography.

The major oceanic divisions – listed below in descending order of area and volume – are defined in part by the continents, various archipelagos, and other criteria.

#	Ocean	Location	Area (km^2) (%)	Volume (km^3) (%)	Avg. depth (m)	Coastline (km)
1	Pacific Ocean	Separates Asia and Oceania from the Americas	168,723,000 *46.6*	669,880,000 *50.1*	3,970	135,663
2	Atlantic Ocean	Separates the Americas from Eurasia and Africa	85,133,000 *23.5*	310,410,900 *23.3*	3,646	111,866
3	Indian Ocean	Washes upon southern Asia and separates Africa and Australia	70,560,000 *19.5*	264,000,000 *19.8*	3,741	66,526
4	Southern Ocean	Sometimes considered an extension of the Pacific, Atlantic and Indian Oceans,which encircles Antarctica	21,960,000 *6.1*	71,800,000 *5.4*	3,270	17,968
5	Arctic Ocean	Sometimes considered a sea or estuary of the Atlantic, which covers much of the Arctic and washes upon northern North America and Eurasia	15,558,000 *4.3*	18,750,000 *1.4*	1,205	45,389
Total – World Ocean			**361,900,000** ***100***	**1,335,000,000** ***100***	**3,688**	**377,412**

NB: Volume, area, and average depth figures include NOAA ETOPO1 figures for marginal South China Sea. Sources: Encyclopedia of Earth, International Hydrographic Organization, *Regional Oceanography: an Introduction* (Tomczak, 2005),Encyclopædia Britannica, and the International Telecommunication Union.

Oceans are fringed by smaller, adjoining bodies of water such as seas, gulfs, bays, bights, and straits.

Global System

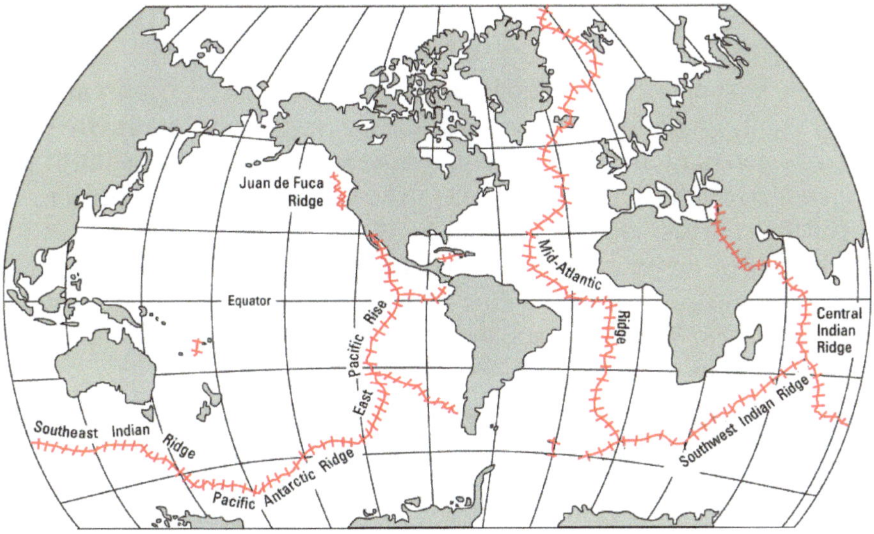

World Distribution of Mid-Oceanic Ridges; USGS

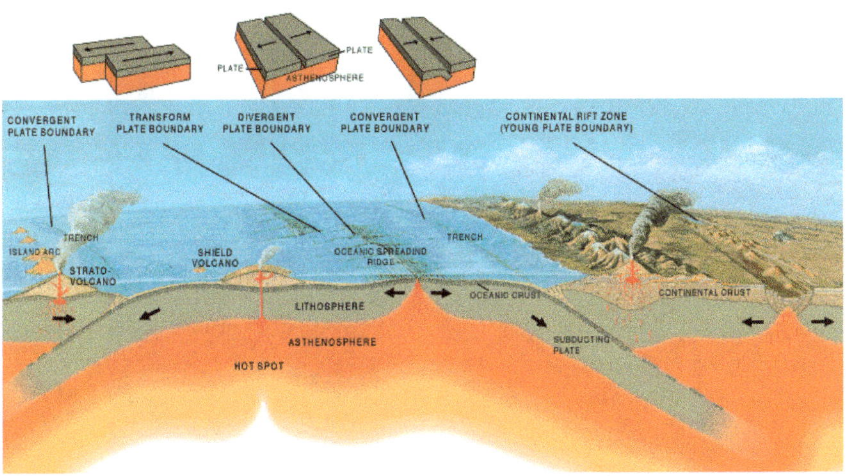

Three main types of plate boundaries.

The Mid-Oceanic Ridge of the World are connected and form *the* Ocean Ridge, a single global mid-oceanic ridge system that is part of every ocean, making it the longest mountain range in the world. The continuous mountain range is 65,000 km (40,400 mi) long (several times longer than the Andes, the longest continental mountain range), and the total length of the oceanic ridge system is 80,000 km (49,700 mi) long.

Physical Properties

The total mass of the hydrosphere is about 1.4 quintillion metric tons (1.4×10^{18} long tons or 1.5×10^{18} short tons), which is about 0.023% of Earth's total mass. Less than 3% is freshwater; the rest is saltwater, almost all of which is in the ocean. The area of the World Ocean is about 361.9 million square kilometers (139.7 million square miles), which covers about 70.9% of Earth's surface, and its volume is approximately 1.335 billion cubic kilometers (320.3 million cubic miles). This can be thought of as a cube of water with an edge length of 1,101 kilometers (684 mi). Its average depth is about 3,688 meters (12,100 ft), and its maximum depth is 10,994 meters (6.831 mi) at the Mariana Trench. Nearly half of the world's marine waters are over 3,000 meters (9,800 ft) deep. The vast expanses of deep ocean (anything below 200 meters or 660 feet) cover about 66% of Earth's surface. This does not include seas not connected to the World Ocean, such as the Caspian Sea.

The bluish color of water is a composite of several contributing agents. Prominent contributors include dissolved organic matter and chlorophyll.

Mariners and other seafarers have reported that the ocean often emits a visible glow which extends for miles at night. In 2005, scientists announced that for the first time, they had obtained photographic evidence of this glow. It is most likely caused by bioluminescence.

Oceanic Zones

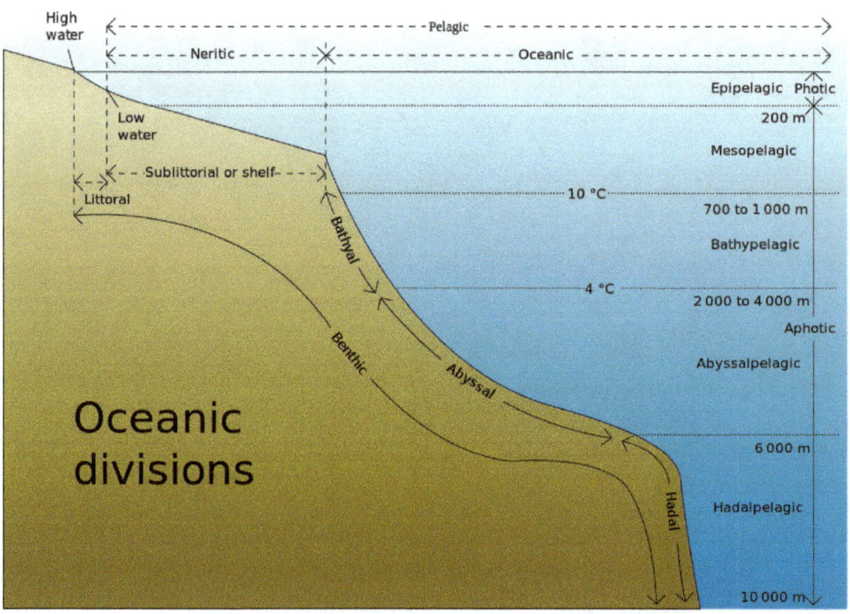

The major oceanic zones, based on depth and biophysical conditions

Oceanographers divide the ocean into different zones by physical and biological conditions. The pelagic zone includes all open ocean regions, and can be divided into further regions categorized by depth and light abundance. The photic zone includes the oceans from the surface to a depth of 200 m; it is the region where photosynthesis can occur and is, therefore, the most biodiverse. Because plants require photosynthesis, life found deeper than the photic zone must either rely on material sinking from above (see marine snow) or find another energy source. Hydrothermal vents are the primary source of energy in what is known as the aphotic zone (depths exceeding 200 m). The pelagic part of the photic zone is known as the epipelagic.

The pelagic part of the aphotic zone can be further divided into vertical regions according to temperature. The mesopelagic is the uppermost region. Its lowermost boundary is at a thermocline of 12 °C (54 °F), which, in the tropics generally lies at 700–1,000 meters (2,300–3,300 ft). Next is the bathypelagic lying between 10 and 4 °C (50 and 39 °F), typically between 700–1,000 meters (2,300–3,300 ft) and 2,000–4,000 meters (6,600–13,100 ft), lying along the top of the abyssal plain is the abyssopelagic, whose lower boundary lies at about 6,000 meters (20,000 ft). The last zone includes the deep oceanic trench, and is known as the hadalpelagic. This lies between 6,000–11,000 meters (20,000–36,000 ft) and is the deepest oceanic zone.

The benthic zones are aphotic and correspond to the three deepest zones of the deep-sea. The bathyal zone covers the continental slope down to about 4,000 meters (13,000 ft). The abyssal zone covers the abyssal plains between 4,000 and 6,000 m. Lastly, the hadal zone corresponds to the hadalpelagic zone, which is found in oceanic trenches.

The pelagic zone can be further subdivided into two subregions: the neritic zone and the oceanic zone. The neritic zone encompasses the water mass directly above the continental shelves whereas the oceanic zone includes all the completely open water.

In contrast, the littoral zone covers the region between low and high tide and represents the transitional area between marine and terrestrial conditions. It is also known as the intertidal zone because it is the area where tide level affects the conditions of the region.

The ocean can be divided into three density zones: the surface zone, the pycnocline, and the deep zone. The surface zone, also called the mixed layer, refers to the uppermost density zone of the ocean. Temperature and salinity are relatively constant with depth in this zone due to currents and wave action. The surface zone contains ocean water that is in contact with the atmosphere and within the photic zone. The surface zone has the ocean's least dense water and represents approximately 2% of the total volume of ocean water. The surface zone usually ranges between depths of 500 feet to 3,300 feet below ocean surface, but this can vary a great deal. In some cases, the surface zone can be entirely non-existent. The surface zone is typically thicker in the tropics than in regions of higher latitude. The transition to colder, denser water is more abrupt in the tropics than in regions of higher latitudes. The pycnocline refers to a zone wherein density substantially increases with depth due primarily to decreases in temperature. The pycnocline effectively separates the lower-density surface zone above from the higher-density deep zone below. The pycnocline represents approximately 18% of the total volume of ocean water. The deep zone refers to the lowermost density zone of the ocean. The deep zone usually begins at depths below 3,300 feet in mid-latitudes. The deep zone undergoes negligible changes in water density with depth. The deep zone represents approximately 80% of the total volume of ocean water. The deep zone contains relatively colder and stable water.

If a zone undergoes dramatic changes in temperature with depth, it contains a thermocline. The tropical thermocline is typically deeper than the thermocline at higher latitudes. Polar waters, which receive relatively little solar energy, are not stratified by temperature and generally lack a thermocline because surface water at polar latitudes are nearly as cold as water at greater depths. Below the thermocline, water is very cold, ranging from −1 °C to 3 °C. Because this deep and cold layer contains the bulk of ocean water, the average temperature of the world ocean is 3.9 °C. If a zone undergoes dramatic changes in salinity with depth, it contains a halocline. If a zone undergoes a strong, vertical chemistry gradient with depth, it contains a chemocline.

The halocline often coincides with the thermocline, and the combination produces a pronounced pycnocline.

Exploration

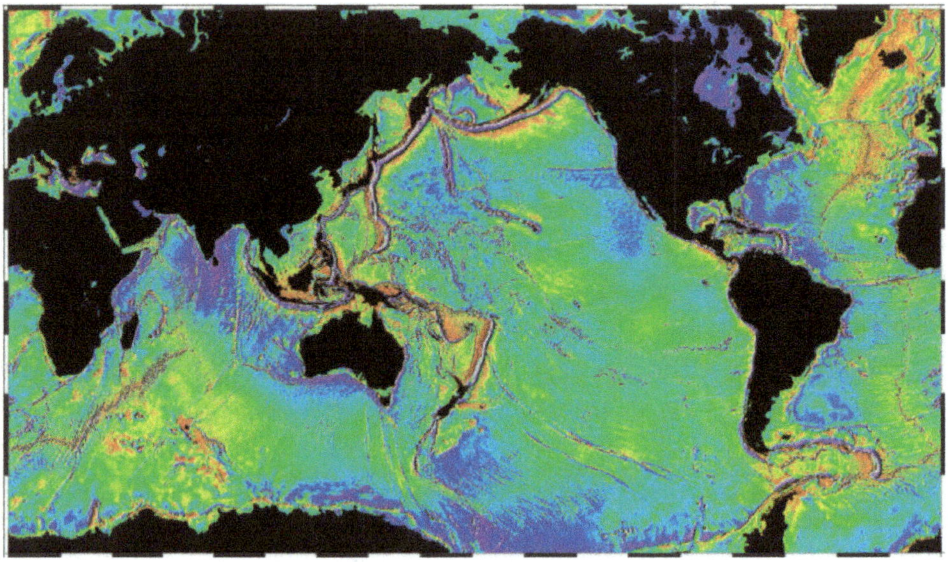

Map of large underwater features (1995, NOAA)

Ocean travel by boat dates back to prehistoric times, but only in modern times has extensive underwater travel become possible.

The deepest point in the ocean is the Mariana Trench, located in the Pacific Ocean near the Northern Mariana Islands. Its maximum depth has been estimated to be 10,971 meters (35,994 ft) (plus or minus 11 meters; see the Mariana Trench article for discussion of the various estimates of the maximum depth.) The British naval vessel *Challenger II* surveyed the trench in 1951 and named the deepest part of the trench the "Challenger Deep". In 1960, the Trieste successfully reached the bottom of the trench, manned by a crew of two men.

Oceanic Maritime Currents

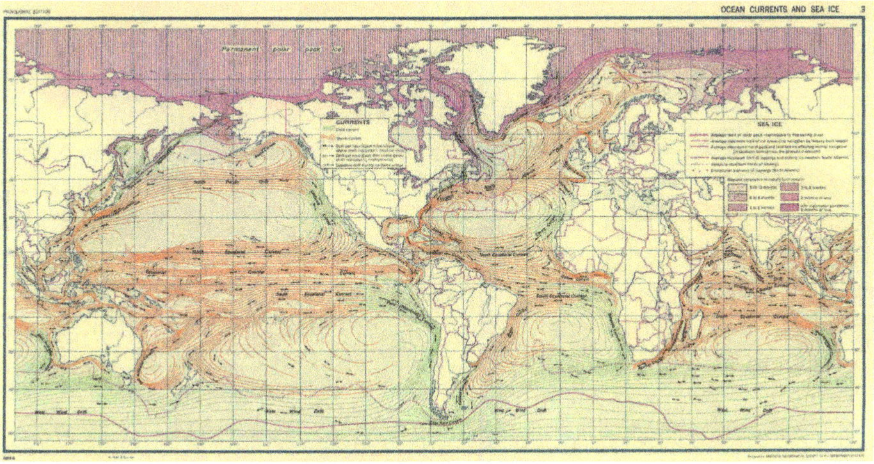

Oceanic currents in 1943.

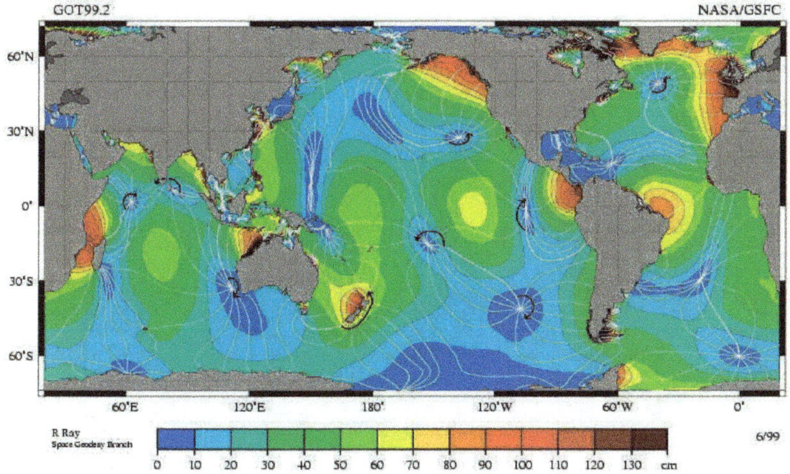

Amphidromic points showing the direction of tides per incrementation periods along with resonating directions of wavelength movements.

Oceanic maritime currents have different origins. Tidal currents are in phase with the tide, hence are quasiperiodic, they may fomulate various knots in certain places, most notably around headlands. Non periodic currents have for origin the waves, wind and different densities.

The wind and waves create surface currents (designated as " drift currents "). These currents can decompose in one quasi permanent current (which varies within the hourly scale) and one movement of Stokes drift under the effect of rapid waves movement (at the echelon of a couple of seconds).).The quasi permanent current is accelerated by the breaking of waves, and in a lesser governing effect, by the friction of the wind on the surface.

This acceleration of the current takes place in the direction of waves and dominant wind. Accordingly, when the sea depth increases, the rotation of the earth changes the direction of currents, in proportion with the increase of depth while friction lowers their speed. At a certain sea depth, the current changes direction and is seen inverted in the opposite direction with speed current becoming nul: known as the Ekman spiral. The influence of these currents is mainly experienced at the mixed layer of the ocean surface, often from 400 to 800 meters of maximum depth. These currents can considerably alter, change and are dependent on the various yearly seasons. If the mixed layer is less thick (10 to 20 meters), the quasi permanent current at the surface adopts an extreme oblique direction in relation to the direction of the wind, becoming virtually homogeneous, until the Thermocline.

In the deep however, maritime currents are caused by the temperature gradients and the salinity between water density masses.

In Littoral zones, Breaking wave is so intense and the depth measurement so low, that maritime currents reach often 1 to 2 knots.

Climate

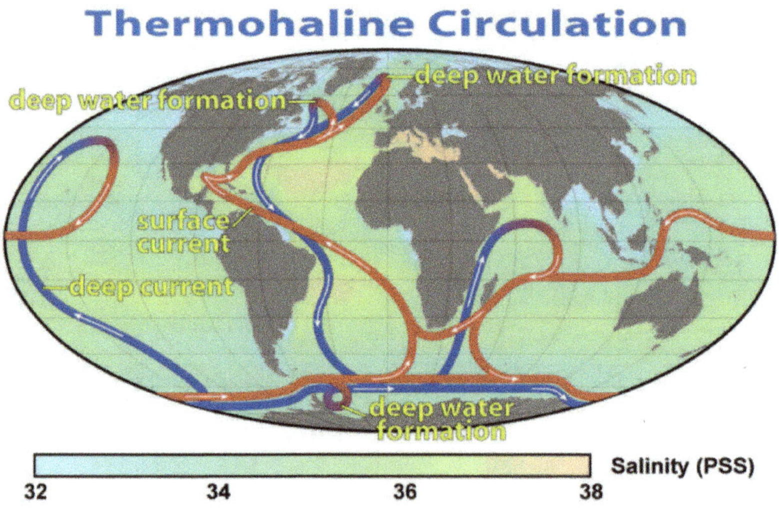

A map of the global thermohaline circulation; blue represent deep-water currents, whereas red represent surface currents

Ocean currents greatly affect Earth's climate by transferring heat from the tropics to the polar regions. Transferring warm or cold air and precipitation to coastal regions, winds may carry them inland. Surface heat and freshwater fluxes create global density gradients that drive the thermohaline circulation part of large-scale ocean circulation.

It plays an important role in supplying heat to the polar regions, and thus in sea ice regulation. Changes in the thermohaline circulation are thought to have significant impacts on Earth's energy budget. In so far as the thermohaline circulation governs the rate at which deep waters reach the surface, it may also significantly influence atmospheric carbon dioxide concentrations.

For a discussion of the possibilities of changes to the thermohaline circulation under global warming, see shutdown of thermohaline circulation.

It is often stated that the thermohaline circulation is the primary reason that the climate of Western Europe is so temperate. An alternate hypothesis claims that this is largely incorrect, and that Europe is warm mostly because it lies downwind of an ocean basin, and because atmospheric waves bring warm air north from the subtropics.

The Antarctic Circumpolar Current encircles that continent, influencing the area's climate and connecting currents in several oceans.

One of the most dramatic forms of weather occurs over the oceans: tropical cyclones (also called "typhoons" and "hurricanes" depending upon where the system forms).

Biology

The ocean has a significant effect on the biosphere. Oceanic evaporation, as a phase of the water cycle, is the source of most rainfall, and ocean temperatures determine climate and wind patterns that affect life on land. Life within the ocean evolved 3 billion years prior to life on land. Both the depth and the distance from shore strongly influence the biodiversity of the plants and animals present in each region.

Lifeforms native to the ocean include:

- Fish;
- Radiata, such as jellyfish (Cnidaria);
- Cetacea, such as whales, dolphins, and porpoises;
- Cephalopods, such as octopus and squid;
- Crustaceans, such as lobsters, shrimp, and krill;
- Marine worms;
- Plankton; and
- Echinoderms, such as brittle stars, starfish, sea cucumbers, and sand dollars.

In addition, many land animals have adapted to living a major part of their life on the oceans. For instance, seabirds are a diverse group of birds that have adapted to a life mainly on the oceans. They feed on marine animals and spend most of their lifetime on water, many only going on land for breeding. Other birds that have adapted to oceans as their living space are penguins, seagulls and pelicans. Seven species of turtles, the sea turtles, also spend most of their time in the oceans.

Gases

Characteristics of Oceanic Gases

Gas	Concentration of Seawater, by Mass (in parts per million), for whole Ocean	% Dissolved Gas, by Volume, in Seawater at Ocean Surface
Carbon dioxide (CO_2)	64 to 107	15%
Nitrogen (N_2)	10 to 18	48%
Oxygen (O_2)	0 to 13	36%

Solubility of Oceanic Gases (in terms of mL/L) with Temperature at salinity of 33‰ and atmospheric pressure

Temperature	O_2	CO_2	N_2
0 °C	8.14	8,700	14.47
10 °C	6.42	8,030	11.59
20 °C	5.26	7,350	9.65
30 °C	4.41	6,600	8.26

Surface

Generalized characteristics of ocean surface by latitude

Characteristic	Oceanic waters in polar regions	Oceanic waters in temperate regions	Oceanic waters in tropical regions
Precipitation vs. evaporation	P > E	P > E	E > P
Sea surface temperature in winter	−2 °C	5 to 20 °C	20 to 25 °C
Average salinity	28‰ to 32‰	35‰	35‰ to 37‰
Annual variation of air temperature	≤ 40ªC	10 °C	< 5 °C
Annual variation of water temperature	< 5ªC	10 °C	< 5 °C

Mixing Time

The *residence time* is the amount of an element in the ocean divided by the rate at which that element is added to (or removed from) the ocean.

The mean oceanic mixing time is thought to be approximately 1,600 years. If a given element in the ocean stays in the ocean, on average, longer than the oceanic mixing time, then that element is assumed to be homogeneously spread throughout the ocean. As a result, because the major salts have a residence time that is longer than 1,600 years, the ratio of major salts is thought to be unchanging across the ocean. This constant ratio is often referred to as Forchhammer's principle or the principle of constant proportions.

Mean oceanic residence time for various constituents

Constituent	Residence time (in years)
Iron (Fe)	200
Aluminum (Al)	600
Manganese (Mn)	1,300
Water (H_2O)	4,100
Silicon (Si)	20,000
Carbonate (CO_3^{2-})	110,000
Calcium (Ca^{2+})	1,000,000
Sulfate (SO_4^{2-})	11,000,000
Potassium (K^+)	12,000,000
Magnesium (Mg^{2+})	13,000,000
Sodium (Na^+)	68,000,000
Chloride (Cl^-)	100,000,000

Salinity

A zone of rapid salinity increase with depth is called a halocline. The temperature of maximum density of seawater decreases as its salt content increases. Freezing temperature of water decreases with salinity, and boiling temperature of water increases with salinity. Typical seawater freezes at around −1.9 °C at atmospheric pressure. If precipitation exceeds evaporation, as is the case in polar and temperate regions, salinity will be lower. If evaporation exceeds precipitation, as is the case in tropical regions, salinity will be higher. Thus, oceanic waters in polar regions have lower salinity content than oceanic waters in temperate and tropical regions.

Salinity can be calculated using the chlorinity, which is a measure of the total mass of halogen ions (includes fluorine, chlorine, bromine, and iodine) in seawater. By international agreement, the following formula is used to determine salinity:

Salinity (in ‰)=1.80655 x Chlorinity (in ‰)

The average chlorinity is about 19.2‰, and, thus, the average salinity is around 34.7‰

Absorption of Light

Absorption of light in different wavelengths by ocean

Color: Wavelength (nm)	Depth wherein 99 percent of wavelength is absorbed (in meters)	Percent absorbed in 1 meter of water
Ultraviolet (UV): 310	31	14.0
Violet (V): 400	107	4.2
Blue (B): 475	254	1.8
Green (G): 525	113	4.0
Yellow (Y): 575	51	8.7
Orange (O): 600	25	16.7
Red (R): 725	4	71.0
Infrared (IR): 800	3	82.0

Economic Value

The oceans are essential to transportation. This is because most of the world's goods move by ship between the world's seaports.Oceans are also the major supply source for the fishing industry. Some of the major harvests are shrimp, fish, crabs, and lobster.

Waves and Swell

The motions of the ocean surface, known as undulations or *waves*, are the partial and alternate rising and falling of the ocean surface. The series of mechanical waves that propagate along the interface between water and air is called swell.

Extraterrestrial Oceans

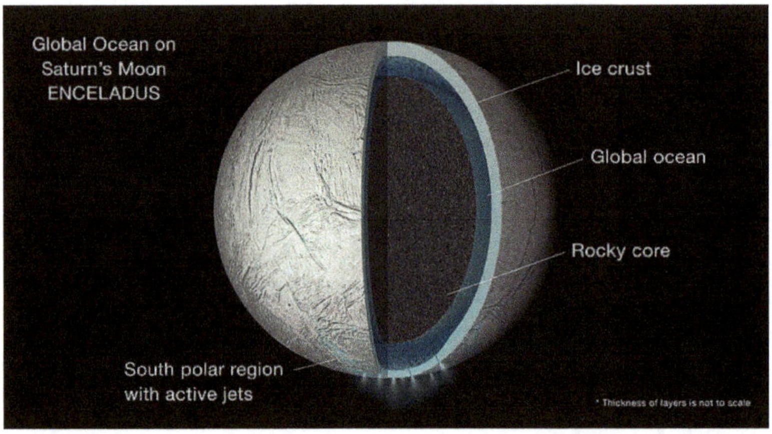

Artist's conception of subsurface ocean of Enceladus confirmed April 3, 2014.

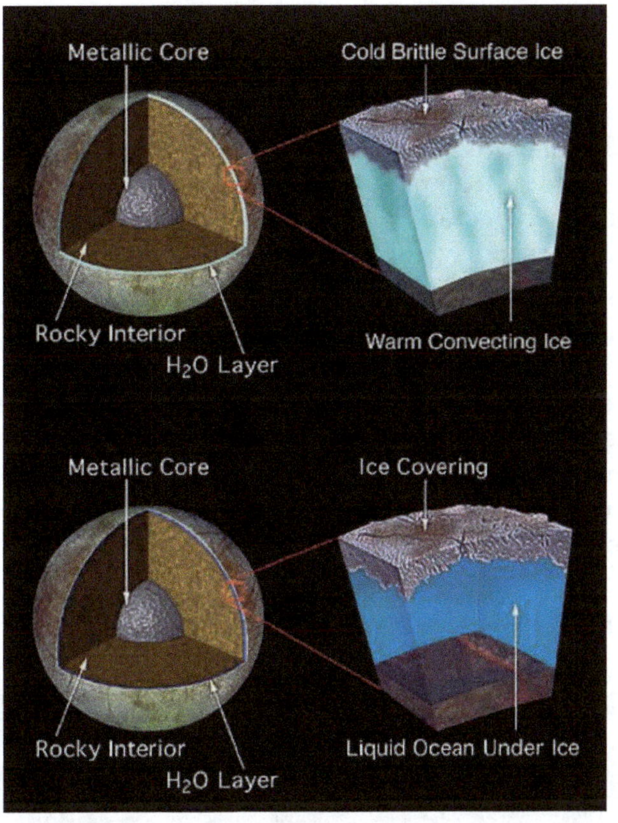

Two models for the composition of Europa predict a large subsurface ocean of liquid water. Similar models have been proposed for other celestial bodies in the Solar System.

Although Earth is the only known planet with large stable bodies of liquid water on its surface and the only one in the Solar System, other celestial bodies are thought to have large oceans.

Planets

The gas giants, Jupiter and Saturn, are thought to lack surfaces and instead have a stratum of liquid hydrogen, however their planetary geology is not well understood. The possibility of the ice giants Uranus and Neptune having hot, highly compressed, supercritical water under their thick atmospheres has been hypothesised. Although their composition is still not fully understood, a 2006 study by Wiktorowicz and Ingersall ruled out the possibility of such a water "ocean" existing on Neptune, though some studies have suggested that exotic oceans of liquid diamond are possible.

The Mars ocean hypothesis suggests that nearly a third of the surface of Mars was once covered by water, though the water on Mars is no longer oceanic (much of it residing in the ice caps). The possibility continues to be studied along with reasons for their apparent disappearance. Astronomers think that Venus had liquid water and perhaps oceans in its very early history. If they existed, all later vanished via resurfacing.

Natural Satellites

A global layer of liquid water thick enough to decouple the crust from the mantle is thought to be present on the natural satellites Titan, Europa, Enceladus and, with less certainty, Callisto, Ganymede and Triton. A magma ocean is thought to be present on Io. Geysers have been found on Saturn's moon Enceladus, possibly originating from about 10 kilometers (6.2 mi) deep ocean beneath an ice shell.Other icy moons may also have internal oceans, or may once have had internal oceans that have now frozen.

Large bodies of liquid hydrocarbons are thought to be present on the surface of Titan, although they are not large enough to be considered oceans and are sometimes referred to as *lakes* or seas. The Cassini–Huygens space mission initially discovered only what appeared to be dry lakebeds and empty river channels, suggesting that Titan had lost what surface liquids it might have had. Cassini's more recent fly-by of Titan offers radar images that strongly suggest hydrocarbon lakes exist near the colder polar regions. Titan is thought to have a subsurface liquid-water ocean under the ice and hydrocarbon mix that forms its outer crust.

Planets and Trans-Neptunian Objects

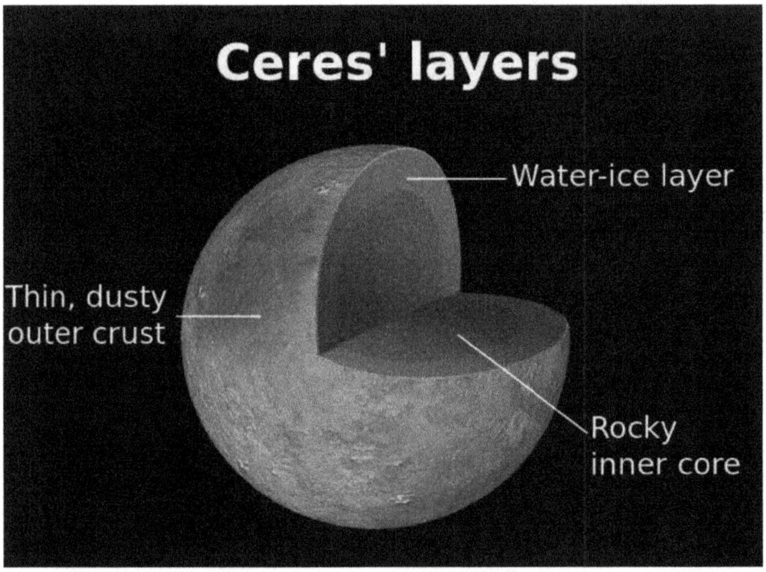

Diagram showing a possible internal structure of Ceres

Ceres appears to be differentiated into a rocky core and icy mantle and may harbour a liquid-water ocean under its surface.

Not enough is known of the larger trans-Neptunian objects to determine whether they are differentiated bodies capable of supporting oceans, although models of radioactive decay suggest that Pluto, Eris, Sedna, and Orcus have oceans beneath solid icy crusts approximately 100 to 180 km thick.

Extrasolar

Rendering of a hypothetical large extrasolar moon with surface liquid-water oceans

Some planets and natural satellites outside the Solar System are likely to have oceans, including possible water ocean planets similar to Earth in the habitable zone or "liquid-water belt". The detection of oceans, even through the spectroscopy method, however is likely extremely difficult and inconclusive.

Theoretical models have been used to predict with high probability that GJ 1214 b, detected by transit, is composed of exotic form of ice VII, making up 75% of its mass, making it an ocean planet.

Other possible candidates are merely speculated based on their mass and position in the habitable zone include planet though little is actually known of their composition. Some scientists speculate Kepler-22b may be an "ocean-like" planet. Models have been proposed for Gliese 581 d that could include surface oceans. Gliese 436 b is speculated to have an ocean of "hot ice". Exomoons orbiting planets, particularly gas giants within their parent star's habitable zone may theoretically have surface oceans.

Terrestrial planets will acquire water during their accretion, some of which will be buried in the magma ocean but most of it will go into a steam atmosphere, and when the atmosphere cools it will collapse on to the surface forming an ocean. There will also be outgassing of water from the mantle as the magma solidifies—this will happen even for planets with a low percentage of their mass composed of water, so "super-Earth exoplanets may be expected to commonly produce water oceans within tens to hundreds of millions of years of their last major accretionary impact."

Non-Water Surface Liquids

Oceans, seas, lakes and other bodies of liquids can be composed of liquids other than water, for example the hydrocarbon lakes on Titan. The possibility of seas of nitrogen on Triton was also considered but ruled out. There is evidence that the icy surfaces of the moons Ganymede, Callisto, Europa, Titan and Enceladus are shells floating on oceans of very dense liquid water or water–ammonia. Earth is often called *the* ocean planet because it is 70% covered in water. Extrasolar terrestrial planets that are extremely close to their parent star will be tidally locked and so one half of the planet will be a magma ocean. It is also possible that terrestrial planets had magma oceans at some point during their formation as a result of giant impacts. Hot Neptunes close to their star could lose their atmospheres via hydrodynamic escape, leaving behind their cores with various liquids on the surface. Where there are suitable temperatures and pressures, volatile chemicals that might exist as liquids in abundant quantities on planets include ammonia, argon, carbon disulfide, ethane, hydrazine, hydrogen, hydrogen cyanide, hydrogen sulfide, methane, neon, nitrogen, nitric oxide, phosphine, silane, sulfuric acid, and water.

Supercritical fluids, although not liquids, do share various properties with liquids. Underneath the thick atmospheres of the planets Uranus and Neptune, it is expected that these planets are composed of oceans of hot high-density fluid mixtures of water, ammonia and other volatiles. The gaseous outer layers of Jupiter and Saturn transition smoothly into oceans of supercritical hydrogen. The atmosphere of Venus is 96.5% carbon dioxide, which is a supercritical fluid at its surface.

ORIGIN OF WATER ON EARTH

The origin of water on Earth, or the reason that there is clearly more liquid water on Earth than on the other rocky planets of the Solar System, is not completely understood. There exist numerous more or less mutually compatible hypotheses as to how water may have accumulated on Earth's surface over the past 4.5 billion years in sufficient quantity to form oceans.

Water covers about 71% of Earth's surface

Possible Sources

Extraplanetary Sources

Comets, trans-Neptunian objects or water-rich meteoroids (protoplanets) from the outer reaches of the asteroid belt colliding with Earth may have brought water to the world's oceans. Measurements of the ratio of the hydrogen isotopes deuterium and protium point to asteroids, since similar percentage impurities in carbon-rich chondrites were found in oceanic water, whereas previous measurement of the isotopes' concentrations in comets and trans-Neptunian objects correspond only slightly to water on Earth.

Planetesimals heated by the decay of aluminium. This could cause water to rise to the surface. Recent studies suggest that water with similar deuterium-to-hydrogen ratio was already available at the time of Earth's formation, as evidenced in ancient "eucrites" meteorites originating from the asteroid Vesta.

That Earth's water originated purely from comets is implausible, since a result of measurements of the isotope ratios of deuterium to protium (D/H ratio) in the four comets Halley, Hyakutake, Hale-Bopp, and 67P/Churyumov–Gerasimenko, by researchers such as David Jewitt, is approximately double that of oceanic water. What is however unclear is whether these comets are representative of those from the Kuiper Belt. According to Alessandro Morbidelli, the largest part of today's water comes from protoplanets formed in the outer asteroid belt that plunged towards Earth, as indicated by the D/H proportions in carbon-rich chondrites. The water in carbon-rich chondrites point to a similar D/H ratio as oceanic water. Nevertheless, mechanisms have been proposed to suggest that the D/H-ratio of oceanic water may have increased significantly throughout Earth's history. Such a proposal is consistent with the possibility that a significant amount of the water on Earth was already present during the planet's early evolution.

Recent measurements of the chemical composition of Moon rocks suggest that Earth was born with its water already present. Investigating lunar samples carried to Earth by the Apollo 15 and 17 missions found a deuterium-to-hydrogen ratio that matched the isotopic ratio in carbonaceous chondrites. The ratio is also similar to that found in water on Earth. The findings suggest a common source of water for both objects. This supports a theory that Jupiter temporarily migrated into the inner Solar System, destabilizing the orbits of water-rich carbonaceous chondrites. As a result, some of the bodies could have fallen inwards and become part of the raw material for making Earth and its neighbors. The discovery of water vapor out-gassing from Ceres provides related information on water-ice content of the asteroid belt.

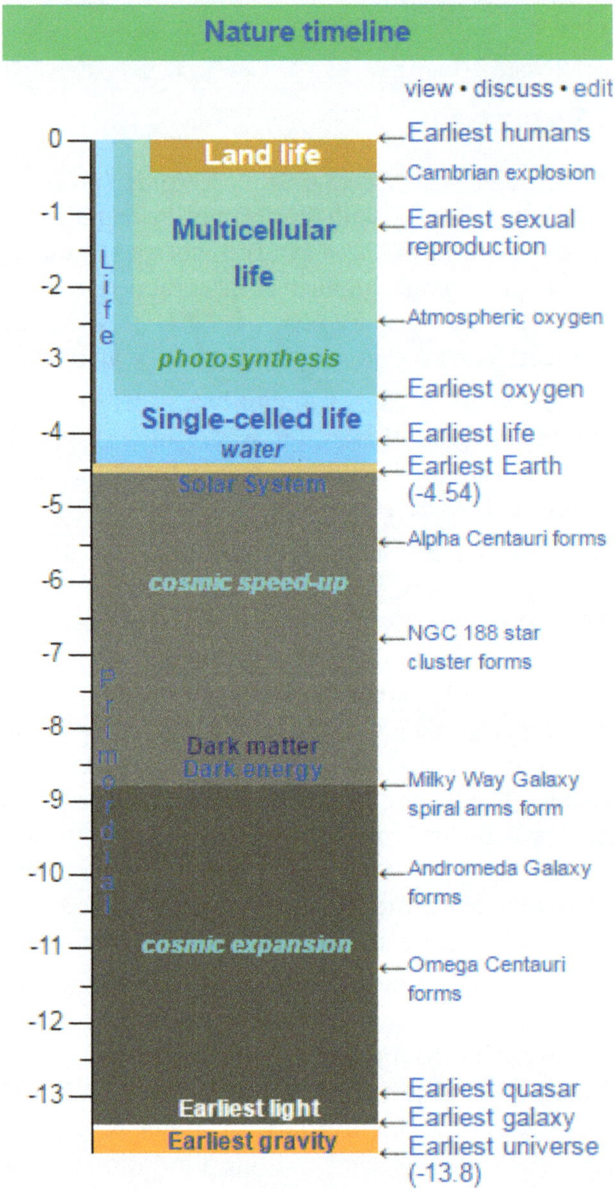

Axis scale: billions of years.

Internal Sources

Gradual leakage of water stored in hydrate minerals of Earth's rocks could have formed a portion of its water. Water may also have come from volcanism: water vapor in the atmosphere that originated in volcanic eruptions may have condensed to form rain, slowly filling Earth's oceanic basins.

Water in the Development of Earth

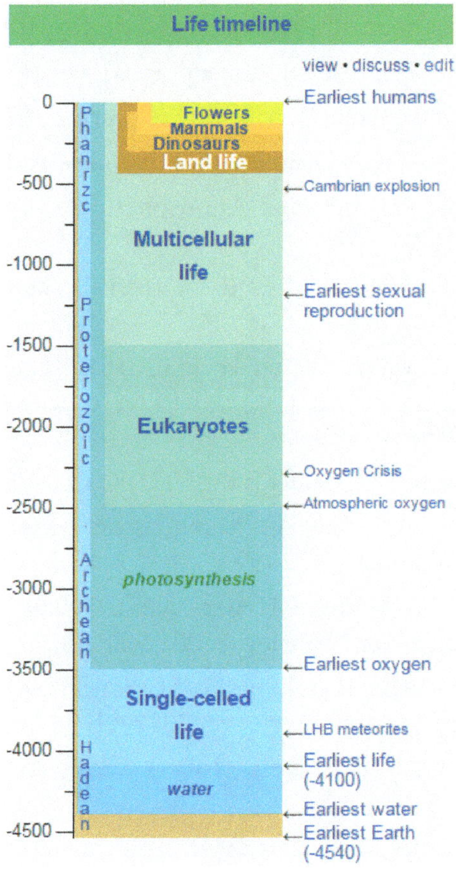

Axis scale: millions of years.

Role of Organisms

Some terrestrial water may have had a biochemical origin, during the Great Oxygenation Event, via redox reactions and photosynthesis.

In the early 1930s, Cornelis Van Niel discovered that sulfide-dependent chemoautotrophic bacteria (purple sulfur bacteria) fix carbon and synthesize water as a byproduct of a photosynthetic pathway using hydrogen sulfide and carbon dioxide:

$$CO_2 + 2H_2S \rightarrow CH_2O + H_2O + 2S$$

Few modern organisms use this method of photosynthesis, making their water contribution negligible. But on the hydrogen-sulfide-rich and oxygen-poor early Earth, a small but significant portion of Earth's water may have been synthesized biochemically through this pathway.

REFERENCES

1. Perseus Digital Library. Retrieved May 17, 2012.
2. "WordNet Search — ocean". Princeton University. Retrieved February 21, 2012.
3. "ocean, n". Oxford English Dictionary. Retrieved February 5, 2012.
4. "ocean". Merriam-Webster. Retrieved February 6, 2012.
5. "WordNet Search — sea". Princeton University. Retrieved February 21, 2012.
6. "NOAA – National Oceanic and Atmospheric Administration – Ocean". Noaa.gov. Retrieved 2012-11-08.
7. Qadri, Syed (2003). "Volume of Earth's Oceans". The Physics Factbook. Retrieved 2007-06-07.
8. Charette, Matthew; Smith, Walter H. F. (2010). "The volume of Earth's ocean". Oceanography. 23 (2): 112–114. doi:10.5670/oceanog.2010.51. Retrieved 27 September 2012.
9. "Volumes of the World's Oceans from ETOPO1". NOAA. Retrieved 2015-03-07.
10. Drogin, Bob (August 2, 2009). "Mapping an ocean of species". Los Angeles Times. Retrieved August 18, 2009.
11. "Titan Likely To Have Huge Underground Ocean | Mind Blowing Science". Mindblowingscience.com. Retrieved 2012-11-08.
12. "Ocean-bearing Planets: Looking For Extraterrestrial Life In All The Right Places". Sciencedaily.com. Retrieved 2012-11-08.
13. □κεανός, Henry George Liddell, Robert Scott, A Greek-English Lexicon, at Perseus project
14. "Ocean". Sciencedaily.com. Retrieved 2012-11-08.
15. ""Distribution of land and water on the planet". UN Atlas of the Oceans. External link in |work= (help)
16. Spilhaus, Athelstan F. (July 1942). "Maps of the whole world ocean". 32 (3). American Geographical Society: 431–5. doi:10.2307/210385.
17. "CIA World Factbook". CIA. Retrieved 2015-04-05.
18. "Pacific Ocean". Encyclopedia of Earth. Retrieved 2015-03-07.
19. "Atlantic Ocean". Encyclopedia of Earth. Retrieved 2015-03-07.
20. "Indian Ocean". Encyclopedia of Earth. Retrieved 2015-03-07.
21. "Southern Ocean". Encyclopedia of Earth. Retrieved 2015-03-10.
22. "Limits of Oceans and Seas, 3rd edition" (PDF). International Hydrographic Organization. 1953. Retrieved 7 February 2010.
23. Tomczak, Matthias; Godfrey, J. Stuart (2003). Regional Oceanography: an Introduction (2 ed.). Delhi: Daya Publishing House. ISBN 81-7035-306-8.
24. "'Arctic Ocean' - Encyclopædia Britannica". Retrieved 2012-07-02. As an approximation, the Arctic Ocean may be regarded as an estuary of the Atlantic Ocean.

25. "Arctic Ocean". Encyclopedia of Earth. Retrieved 2015-03-07.
26. "Recommendation ITU-R RS.1624: Sharing between the Earth exploration-satellite (passive) and airborne altimeters in the aeronautical radionavigation service in the band 4 200-4 400 MHz (Question ITU-R 229/7)" (PDF). ITU Radiotelecommunication Sector (ITU-R). Retrieved 2015-04-05. The oceans occupy about 3.35×108km2 of area. There are 377412 km of oceanic coastlines in the world.
27. "What is the longest mountain range on earth?". Ocean Facts. NOAA. Retrieved 17 October 2014.
28. "Scientists map Mariana Trench, deepest known section of ocean in the world". The Telegraph. Telegraph Media Group. 7 December 2011. Retrieved 23 March 2012.
29. Drazen, Jeffrey C. "Deep-Sea Fishes". School of Ocean and Earth Science and Technology, the University of Hawai□i at Mānoa. Retrieved 2007-06-07.
30. Coble, Paula G. (2007). "Marine Optical Biogeochemistry: The Chemistry of Ocean Color". Chemical Reviews. 107: 402–418. doi:10.1021/cr050350.
31. Britt, Robert Roy (October 4, 2005). "Mystery Ocean Glow Confirmed in Satellite Photos".
32. Holladay, April (November 21, 2005). "A glowing sea, courtesy of algae". USA Today.
33. "Sea's eerie glow seen from space". New Scientist. October 5, 2005.
34. Casey, Amy (August 8, 2003). "The Incredible Glowing Algae". NASA Earth Observatory. NASA.
35. Étude de la dérive à la surface sous l'effet du vent, Observation and estimation of Lagrangian, Stokes and Eulerian currents induced by wind and waves at the sea surface, F. Ardhuin, L. Marié, N. Rascle, P. Forget, and A. Roland, 2009: J. Phys. Oceanogr., vol. 39, n° 11, p.2820–2838.
36. Mesure de l'effet de frottement à la surface de la mer, Tangential stress beneath wind-driven air-water interfaces, M. L. Banner and W. L. Peirson, J. Fluid Mech., vol. 364, p. 115–145, 1998.
37. Courants mesurés près de la surface, The drift current from observations made on the bouee laboratoire, Joseph Gonella, 1971: Cahiers Océanographiques, vol. 23, p. 1–15.
38. Seager, R. (2006). "The Source of Europe's Mild Climate". American Scientist. 94 (4): 334. doi:10.1511/2006.60.999.
39. Rhines; Hakkinen (2003). "Is the Oceanic Heat Transport in the North Atlantic Irrelevant to the Climate in Europe?" (PDF). ASOF Newsletter.
40. "Chapter 34: The Biosphere: An Introduction to Earth's Diverse Environment". Biology: Concepts & Connections. section 34.7.
41. "Dissolved Gases other than Carbon Dioxide in Seawater" (PDF). soest.hawaii.edu. Retrieved 2014-05-05.

42. "Dissolved Oxygen and Carbon Dioxide" (PDF). chem.uiuc.edu.
43. Anthoni, Floor. "Composition of seawater". Seafriends.org.nz. Retrieved 2014-05-05.
44. "12.742. Marine Chemistry. Lecture 8. Dissolved Gases and Air-sea exchange" (PDF). Retrieved 2014-05-05.
45. "5.6 Synthesis – AR4 WGI Chapter 5: Observations: Oceanic Climate Change and Sea Level". Ipcc.ch. Retrieved 2014-05-05.
46. "Evaporation minus precipitation, Latitude-Longitude, Annual mean". ERA-40 Atlas. ECMWF. Archived from the original on 2014-02-02.
47. Barry, Roger Graham; Chorley, Richard J. (2003). "Atmosphere, Weather, and Climate".. p. 68.
48. "Ocean Stratification". Eesc.columbia.edu. Retrieved 2014-05-05.
49. Huang, Rui Xin (2010). "Ocean Circulation: Wind-Driven and Thermohaline Processes"..
50. Deser, C.; Alexander, M. A.; Xie, S. P.; Phillips, A. S. (2010). "Sea Surface Temperature Variability: Patterns and Mechanisms" (PDF). Annual Review of Marine Science. 2: 115–43. doi:10.1146/annurev-marine-120408-151453. PMID 21141660.
51. "Chapter 6 – Temperature, Salinity, and Density – Geographical Distribution of Surface Temperature and Salinity". Introduction to Physical Oceanography :. Oceanworld.tamu.edu. 2009-03-23. Retrieved 2014-05-05.
52. "Calculation of residence times in seawater of some important solutes" (PDF). gly.uga.edu.
53. Chester, Roy; Jickells, Tim (2012). Marine Geochemistry. Blackwell Publishing. pp. 225–230. ISBN 978-1-118-34907-6.
54. Chester, Roy; Jickells, Tim (2012). Marine Geochemistry. Blackwell Publishing. ISBN 978-1-118-34907-6.
55. Platt, Jane; Bell, Brian (2014-04-03). "NASA Space Assets Detect Ocean inside Saturn Moon". NASA. Retrieved 2014-04-03.
56. Iess, L.; Stevenson, D.J.; Parisi, M.; Hemingway, D.; et al. (4 April 2014). "The Gravity Field and Interior Structure of Enceladus". Science. 344 (6179): 78–80. doi:10.1126/science.1250551. Retrieved 3 April 2014.
57. Dyches, Preston; Chou, Felcia (7 April 2015). "The Solar System and Beyond is Awash in Water". NASA. Retrieved 8 April 2015.
58. Wiktorowicz, Sloane J.; Ingersoll, Andrew P. (2007). "Liquid water oceans in ice giants". Icarus. 186 (2): 436–447. arXiv:astro-ph/0609723 . Bibcode:2007Icar..186..436W. doi:10.1016/j.icarus.2006.09.003. ISSN 0019-1035.
59. Silvera, Isaac (2010). "Diamond: Molten under pressure". Nature Physics. 6 (1): 9–10. Bibcode:2010NatPh...6....9S. doi:10.1038/nphys1491. ISSN 1745-2473.
60. Clavin, Whitney (May 1, 2014). "Ganymede May Harbor ‹Club Sandwich' of Oceans and Ice". NASA. Jet Propulsion Laboratory. Retrieved 2014-05-01.

61. Vance, Steve; Bouffard, Mathieu; Choukroun, Mathieu; Sotina, Christophe (12 April 2014). "Ganymede's internal structure including thermodynamics of magnesium sulfate oceans in contact with ice". Planetary and Space Science. 96: 62–70. Bibcode:2014P&SS...96...62V. doi:10.1016/j.pss.2014.03.011. Retrieved 2014-05-02.

62. McKinnon, William B.; Kirk, Randolph L. (2007). "Triton". In Lucy Ann Adams McFadden; Lucy-Ann Adams; Paul Robert Weissman; Torrence V. Johnson. Encyclopedia of the Solar System (2nd ed.). Amsterdam; Boston: Academic Press. pp. 483–502. ISBN 978-0-12-088589-3.

63. Ruiz, Javier (December 2003). "Heat flow and depth to a possible internal ocean on Triton". Icarus. 166 (2): 436–439. Bibcode:2003Icar..166..436R. doi:10.1016/j.icarus.2003.09.009.

64. Hussmann, Hauke; Sohl, Frank; Spohn, Tilman (November 2006). "Subsurface oceans and deep interiors of medium-sized outer planet satellites and large trans-neptunian objects" (PDF). Icarus. 185 (1): 258–273. Bibcode:2006Icar..185..258H. doi:10.1016/j.icarus.2006.06.005.

65. McCord, Thomas B. (2005). "Ceres: Evolution and current state". Journal of Geophysical Research. 110 (E5): E05009. Bibcode:2005JGRE..11005009M. doi:10.1029/2004JE002244.

66. Castillo-Rogez, J. C.; McCord, T. B.; Davis, A. G. (2007). "Ceres: evolution and present state" (PDF). Lunar and Planetary Science. XXXVIII: 2006–2007. Retrieved 2009-06-25.

67. "The Inside Story". pluto.jhuapl.edu — NASA New Horizons mission site. Johns Hopkins University Applied Physics Laboratory. 2013. Retrieved 2 August 2013.

68. Aguilar, David A. (2009-12-16). "Astronomers Find Super-Earth Using Amateur, Off-the-Shelf Technology". Harvard-Smithsonian Center for Astrophysics. Retrieved January 23, 2010.

69. Mendez Torres, Abel (2011-12-08). "Updates on Exoplanets during the First Kepler Science Conference". Planetary Habitability Laboratory at UPR Arecibo.

70. Fox, Maggie (May 16, 2007). "Hot "ice" may cover recently discovered planet". Reuters. Retrieved May 18, 2012.

71. Elkins-Tanton (2010). "Formation of Early Water Oceans on Rocky Planets". Astrophysics and Space Science. 332 (2): 359–364. arXiv:1011.2710 . Bibcode:2011Ap&SS.332..359E. doi:10.1007/s10509-010-0535-3.

72. McKinnon, William B.; Kirk, Randolph L. (2007). "Triton". In Lucy Ann Adams McFadden; Lucy-Ann Adams; Paul Robert Weissman; Torrence V. Johnson. Encyclopedia of the Solar System (2nd ed.). Amsterdam; Boston: Academic Press. p. 485. ISBN 978-0-12-088589-3.

73. Coustenis, A.; Lunine, J.; Lebreton, J.; Matson, D.; et al. (2008). "American Geophysical Union, Fall Meeting 2008, abstract #P21A-1346". American Geophysical Union. 21: 1346. Bibcode:2008AGUFM.P21A1346C. the Titan system, rich in organics, containing a vast subsurface ocean of liquid water |contribution=ignored (help)

74. Nimmo, F.; Bills, B. G. (2010). "Shell thickness variations and the long-wavelength topography of Titan". Icarus. 208 (2): 896–904. Bibcode:2010Icar..208..896N. doi:10.1016/j.icarus.2010.02.020. observations can be explained if Titan has a floating, isostatically-compensated ice shell
75. Goldreich, Peter M.; Mitchell, Jonathan L. (2010). "Elastic ice shells of synchronous moons: Implications for cracks on Europa and non-synchronous rotation of Titan". Icarus. 209 (2): 631–638. arXiv:0910.0032 . Bibcode:2010Icar..209..631G. doi:10.1016/j.icarus.2010.04.013. A number of synchronous moons are thought to harbor water oceans beneath their outer ice shells. A subsurface ocean frictionally decouples the shell from the interior
76. "Study of the ice shells and possible subsurface oceans of the Galilean satellites using laser altimeters on board the Europa and Ganymede orbiters JEO and JGO" (PDF). Retrieved 2011-10-14.
77. "Tidal heating and the long-term stability of a subsurface ocean on Enceladus" (PDF). Retrieved 2011-10-14.
78. Hinrichsen, D (2011-10-03). "The ocean planet". People Planet. 7: 6–9. PMID 12349465.
79. "Irrigating Crops with Seawater". Scientific American. August 1998.
80. Schaefer, Laura; Fegley, Bruce, Jr. (2009). "Chemistry of Silicate Atmospheres of Evaporating Super-Earths". The Astrophysical Journal Letters. 703 (2): L113–L117. arXiv:0906.1204 . Bibcode:2009ApJ...703L.113S. doi:10.1088/0004-637X/703/2/L113.
81. Solomatov, V. S. (2000). "Fluid Dynamics of a Terrestrial Magma Ocean" (PDF).
82. Leitner, J.J.; Lammer, H.; Odert, P.; Leitzinger, M.; et al. (2009). "Atmospheric Loss of Sub-Neptune's and Implications for Liquid Phases of Different Solvents on Their Surfaces" (PDF). EPSC Abstracts. 4. EPSC2009-542.
83. Tables 3 and 4 in Bains, William (2004). "Many Chemistries Could Be Used to Build Living Systems" (PDF). Astrobiology.
84. Atreya, S.; Egeler, P.; Baines, K. (2006). "Water-ammonia ionic ocean on Uranus and Neptune?" (PDF). Geophysical Research Abstracts. 8: 05179. Bibcode:2005AGUFM.P11A0088A.
85. Guillot, T. (1999). "A comparison of the interiors of Jupiter and Saturn". Planetary and Space Science. 47 (10–11): 1183–200. arXiv:astro-ph/9907402 . Bibcode:1999P&SS...47.1183G. doi:10.1016/S0032-0633(99)00043-4.
86. Lang, Kenneth R. (2003). "Jupiter: a giant primitive planet". NASA. Retrieved 2007-01-10.
87. "The World Factbook". www.cia.gov. Retrieved 2016-03-17.
88. Altwegg, K.; Balsiger, H.; Bar-Nun, A.; Berthelier, J. J.; Bieler, A.; Bochsler, P.; Briois, C.; Calmonte, U.; Combi, M. (2015-01-23). "67P/Churyumov-Gerasimenko, a Jupiter family comet with a high D/H ratio". Science. 347 (6220): 1261952. doi:10.1126/science.1261952. ISSN 0036-8075. PMID 25501976.
89. Sarafian, Adam R.; Nielsen, Sune G.; Marschall, Horst R.; McCubbin, Francis

M.; Monteleone, Brian D. (2014-10-31). "Early accretion of water in the inner solar system from a carbonaceous chondrite–like source". Science. 346 (6209): 623–626. doi:10.1126/science.1256717. ISSN 0036-8075. PMID 25359971.

90. Alessandro Morbidelli et al. Meteoritics & Planetary Science 35, 2000, S. 1309–1329

91. H. Genda, M. Ikoma, Origin of the Ocean on the Earth: Early Evolution of Water D/H in a Hydrogen-rich Atmosphere. Accessible at http://arxiv.org/abs/0709.2025

92. Cowen, Ron (9 May 2013). "Common source for Earth and Moon water". Nature. doi:10.1038/nature.2013.12963.

93. "Herschel discovers water vapour around dwarf planet Ceres". European Space Agency. Retrieved 10 February 2014.

94. http://live.huffingtonpost.com/r/highlight/scientists-discover-where-earths-water-originated/564ba93d99ec6d09c3000126?source=gravityRR&cps=gravity_5060_-7528146454370360511

95. Drake, Michael J. (April 2005). "Origin of water in the terrestrial planets". Meteoritics & Planetary Science. John Wiley & Sons. 40 (4): 519–527. Bibcode:2005M&PS...40..519D. doi:10.1111/j.1945-5100.2005.tb00960.x.

96. Drake, Michael J.; et al. (August 2005). "Origin of water in the terrestrial planets". Asteroids, Comets, and Meteors (IAU S229). 229th Symposium of the International Astronomical Union. 1. Búzios, Rio de Janeiro, Brazil: Cambridge University Press. pp. 381–394. Bibcode:2006IAUS..229..381D. doi:10.1017/S1743921305006861. ISBN 978-0521852005.

97. "Solar System Exploration: Science & Technology: Science Features: View Feature". Solarsystem.nasa.gov. 2004-04-26. Retrieved 2009-08-20.

98. N. H. Sleep; K. Zahnle & P. S. Neuhoff. "Inaugural Article: Initiation of clement surface conditions on the earliest Earth - Sleep et al. 98 (7): 3666 - Proceedings of the National Academy of Sciences". Pnas.org. Retrieved 2009-08-20.

99. "ANU - Research School of Earth Sciences - ANU College of Science - Harrison". Ses.anu.edu.au. Retrieved 2009-08-20.

100. "ANU - OVC - MEDIA - MEDIA RELEASES - 2005 - NOVEMBER - 181105HARRISONCONTINENTS". Info.anu.edu.au. Retrieved 2009-08-20.

101. "A Cool Early Earth". Geology.wisc.edu. Retrieved 2009-08-20.

102. Chang, Kenneth (2008-12-02). "A New Picture of the Early Earth". The New York Times. Retrieved 2010-05-20.

103. "The oxygenation of the atmosphere and oceans" (PDF), Philosophical Transactions of The Royal Society: Biological Sciences, 29 June 2006

104. van Niel, C.B. (1931). "Photosynthesis of bacteria". Arch. Mikrobiol. 3 (1).

CHAPTER 2

Seabed

Chapter Contents

Seabed is an experimental industrial field which involves extracting submerged minerals and deposits from the sea floor. There are interests both for and against seabed mining, however, the science around the environmental impact of SBM is incomplete and unproven. In addition, there is also new interest to explore Deep Sea Mining (DSM), which requires intensive, destructive processes to retrieve deposits laid down over thousands of years around underwater hot springs or hydrothermal vents in the ocean.

INTRODUCTION

The seabed is the bottom of the ocean.

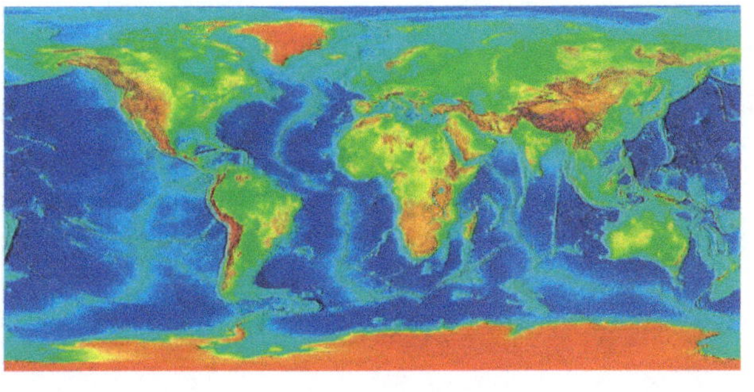

Map showing the underwater topography (bathymetry) of the ocean floor. Like land terrain, the ocean floor has ridges, valleys, plains and volcanoes.

Seabed Structure

Most of the oceans have a common structure, created by common physical phenomena, mainly from tectonic movement, and sediment from various sources. The structure of the oceans, starting with the continents, begins usually with a continental shelf, continues to the continental slope – which is a steep descent into the ocean, until reaching the abyssal plain – a topographic plain, the beginning of the seabed, and its main area. The border between the continental slope and the abyssal plain usually has a more gradual descent, and is called the continental rise, which is caused by sediment cascading down the continental slope.

The mid-ocean ridge, as its name implies, is a mountainous rise through the middle of all the oceans, between the continents. Typically a rift runs along the edge of this ridge. Along tectonic plate edges there are typically oceanic trenches – deep valleys, created by the mantle circulation movement from the mid-ocean mountain ridge to the oceanic trench.

Hotspot volcanic island ridges are created by volcanic activity, erupting periodically, as the tectonic plates pass over a hotspot. In areas with volcanic activity and in the oceanic trenches there are hydrothermal vents – releasing high pressure and extremely hot water and chemicals into the typically freezing water around it.

Deep ocean water is divided into layers or zones, each with typical features of salinity, pressure, temperature and marine life, according to their depth. Lying along the top of the abyssal plain is the abyssal zone, whose lower boundary lies at about 6,000 m (20,000 ft). The hadal zone – which includes the oceanic trenches, lies between 6,000–11,000 metres (20,000–36,000 ft) and is the deepest oceanic zone.

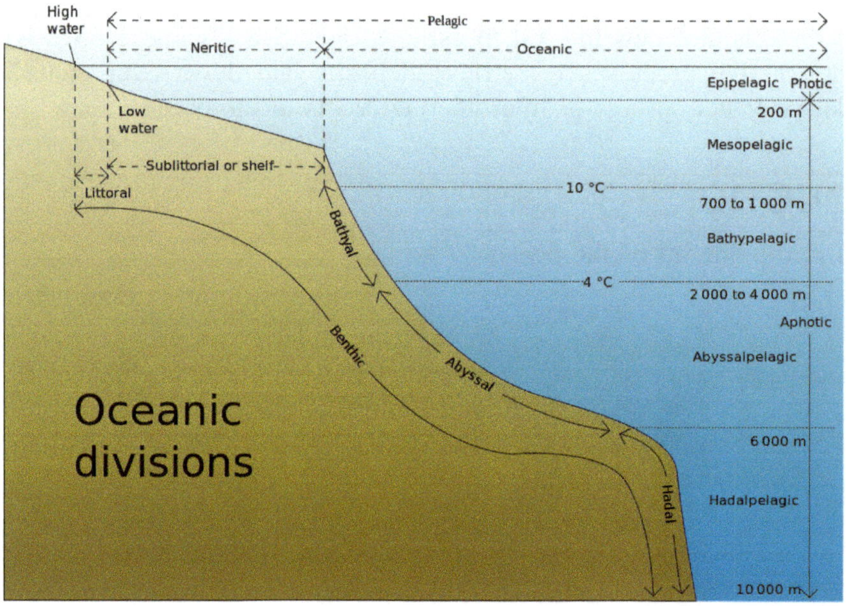

The major oceanic divisions

SEAFLOOR SPREADING

Seafloor spreading is a process that occurs at mid-ocean ridges, where new oceanic crust is formed through volcanic activity and then gradually moves away from the ridge. Seafloor spreading helps explain continental drift in the theory of plate tectonics. When oceanic plates diverge, tensional stress causes fractures to occur in the lithosphere. Basaltic magma rises up the fractures and cools on the ocean floor to form new sea floor. Older rocks will be found farther away from the spreading zone while younger rocks will be found nearer to the spreading zone.

Earlier theories (e.g. by Alfred Wegener and Alexander du Toit) of continental drift were that continents "ploughed" through the sea. The idea that the seafloor itself moves (and also carries the continents with it) as it expands from a central axis was proposed by Harry Hess from Princeton University in the 1960s. The theory is well accepted now, and the phenomenon is known to be caused by convection currents in the asthenosphere, which is ductile, or plastic, and the brittle lithosphere.

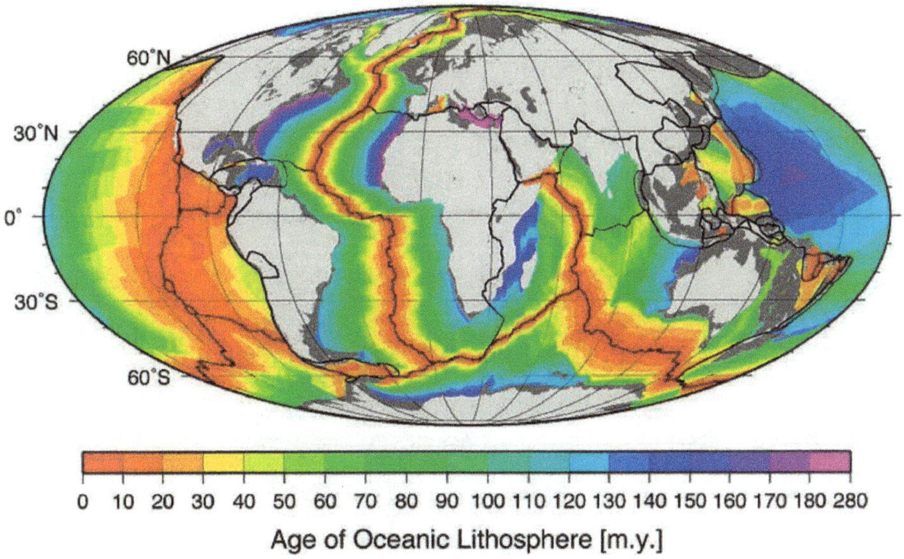

Age of oceanic lithosphere; youngest (red) is along spreading centers.

Incipient Spreading

In the general case, sea floor spreading starts as a rift in a continental land mass, similar to the Red Sea-East Africa Rift System today. The process starts with heating at the base of the continental crust which causes it to become more plastic and less dense. Because less dense objects rise in relation to denser objects, the area being heated becomes a broad dome (see isostasy). As the crust bows upward, fractures occur that gradually grow into rifts. The typical rift system consists of three rift arms at approximately 120 degree angles. These areas are named triple junctions and can be found in several places across the world today. The separated margins of the continents evolve to form passive margins.

Hess' theory was that new seafloor is formed when magma is forced upward toward the surface at a mid-ocean ridge.

If spreading continues past the incipient stage described above, two of the rift arms will open while the third arm stops opening and becomes a 'failed rift'. As the two active rifts continue to open, eventually the continental crust is attenuated as far as it will stretch. At this point basaltic oceanic crust begins to form between the separating continental fragments. When one of the rifts opens into the existing ocean, the rift system is flooded with seawater and becomes a new sea. The Red Sea is an example of a new arm of the sea. The East African rift was thought to be a "failed" arm that was opening somewhat more slowly than the other two arms, but in 2005 the Ethiopian Afar Geophysical Lithospheric Experiment reported that in the Afar region last September, a 60 km fissure opened as wide as eight meters. During this period of initial flooding the new sea is sensitive to changes in climate and eustasy. As a result, the new sea will evaporate (partially or completely) several times before the elevation of the rift valley has been lowered to the point that the sea becomes stable. During this period of evaporation large evaporite deposits will be made in the rift valley. Later these deposits have the potential to become hydrocarbon seals and are of particular interest to petroleum geologists.

Sea floor spreading can stop during the process, but if it continues to the point that the continent is completely severed, then a new ocean basin is created. The Red Sea has not yet completely split Arabia from Africa, but a similar feature can be found on the other side of Africa that has broken completely free. South America once fit into the area of the Niger Delta. The Niger River has formed in the failed rift arm of the triple junction.

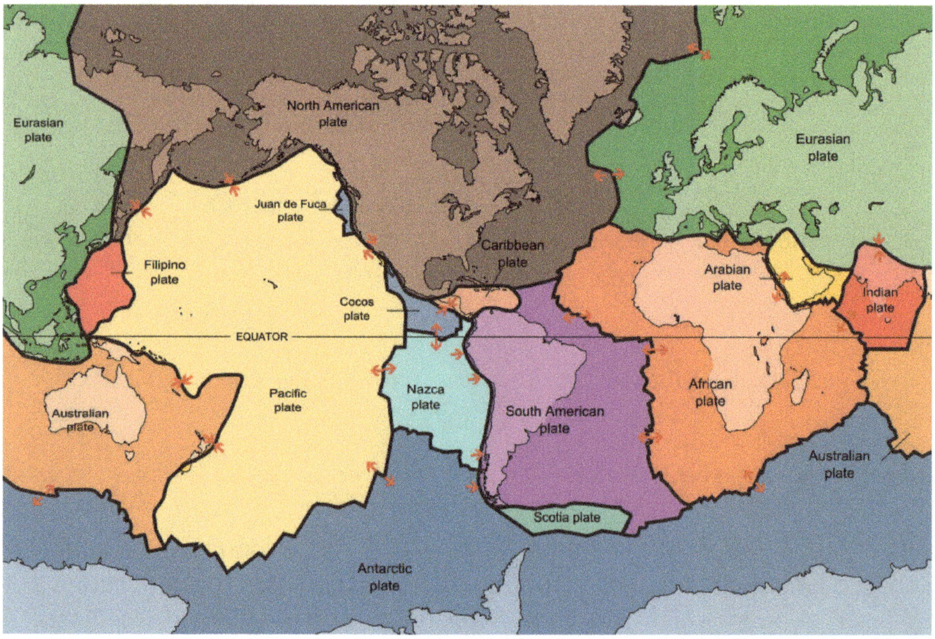

Plates in the crust of the earth, according to the plate tectonics theory

Continued Spreading and Subduction

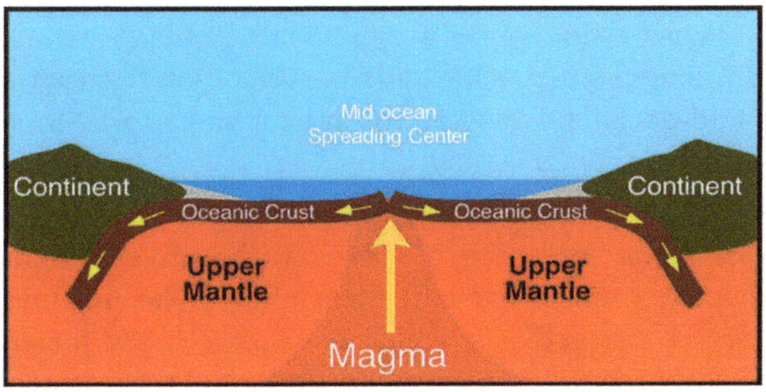

Spreading at a mid-ocean ridge

As new seafloor forms and spreads apart from the mid-ocean ridge it slowly cools over time. Older seafloor is therefore colder than new seafloor, and older oceanic basins deeper than new oceanic basins due to isostasy. If the diameter of the earth remains relatively constant despite the production of new crust, a mechanism must exist by which crust is also destroyed. The destruction of oceanic crust occurs at subduction zones where oceanic crust is forced under either continental crust or oceanic crust. Today, the Atlantic basin is actively spreading at the Mid-Atlantic Ridge. Only a small portion of the oceanic crust produced in the Atlantic is subducted. However, the plates making up the Pacific Ocean are experiencing subduction along many of their boundaries which causes the volcanic activity in what has been termed the Ring of Fire of the Pacific Ocean. The Pacific is also home to one of the world's most active spreading centres (the East Pacific Rise) with spreading rates of up to 13 cm/yr. The Mid-Atlantic Ridge is a "textbook" slow-spreading centre, while the East Pacific Rise is used as an example of fast spreading. The differences in spreading rates affect not only the geometries of the ridges but also the geochemistry of the basalts that are produced.

Since the new oceanic basins are shallower than the old oceanic basins, the total capacity of the world's ocean basins decreases during times of active sea floor spreading. During the opening of the Atlantic Ocean, sea level was so high that a Western Interior Seaway formed across North America from the Gulf of Mexico to the Arctic Ocean.

Debate and Search for Mechanism

At the Mid-Atlantic Ridge (and in other areas), material from the upper mantle rises through the faults between oceanic plates to form new crust as the plates move away from each other, a phenomenon first observed as continental drift. When Alfred Wegener first presented a hypothesis of continental drift in 1912, he suggested that continents ploughed through the ocean crust. This was impossible: oceanic crust is both more dense and more rigid than continental crust. Accordingly, Wegener's theory wasn't taken very seriously, especially in the United States.

Since then, it has been shown that the motion of the continents is linked to seafloor spreading. In the 1960s, the past record of geomagnetic reversals was noticed by observing the magnetic stripe "anomalies" on the ocean floor. This results in broadly evident "stripes" from which the past magnetic field polarity can be inferred by looking at the data gathered from simply towing a magnetometer on the sea surface or from an aircraft. The stripes on one side of the mid-ocean ridge were the mirror image of those on the other side. The seafloor must have originated on the Earth's great fiery welts, like the Mid-Atlantic Ridge and the East Pacific Rise.

The driver for seafloor spreading in plates with active margins is the weight of the cool, dense, subducting slabs that pull them along. The magmatism at the ridge is considered to be "passive upswelling", which is caused by the plates being pulled apart under the weight of their own slabs. This can be thought of as analogous to a rug on a table with little friction: when part of the rug is off of the table, its weight pulls the rest of the rug down with it.

Sea Floor Global Topography: Half-Space Model

To first approximation, sea floor global topography in areas without significant subduction can be estimated by the half-space model. In this model, the seabed height is determined by the oceanic lithosphere temperature, due to thermal expansion. Oceanic lithosphere is continuously formed at a constant rate at the mid-ocean ridges. The source of the lithosphere has a half-plane shape ($x = 0$, $z < 0$) and a constant temperature T_1. Due to its continuous creation, the lithosphere at $x > 0$ is moving away from the ridge at a constant velocity v, which is assumed large compared to other typical scales in the problem. The temperature at the upper boundary of the lithosphere ($z=0$) is a constant $T_0 = 0$. Thus at $x = 0$ the temperature is the Heaviside step function $T_1 \cdot \Theta(-z)$. Finally, we assume the system is at a quasi-steady state, so that the temperature distribution is constant in time, i.e. $T=T(x,z)$.

By calculating in the frame of reference of the moving lithosphere (velocity v), which have spatial coordinate $x' = x-vt$, we may write $T = T(x',z,t)$ and use the heat equation: $\frac{\partial T}{\partial t} = \kappa\nabla^2 T = \kappa\frac{\partial^2 T}{\partial^2 z} + \kappa\frac{\partial^2 T}{\partial^2 x'}$ is where k the thermal diffusivity of the mantle lithosphere.

Since T depends on x' and t only through the combination $x = x' + vt$, we have: $\frac{\partial T}{\partial x'} = \frac{1}{v} \cdot \frac{\partial T}{\partial t}$

Thus: $\frac{\partial T}{\partial t} = \kappa\nabla^2 T = \kappa\frac{\partial^2 T}{\partial^2 z} + \frac{\kappa}{v^2}\frac{\partial^2 T}{\partial^2 t}$

We now use the assumption that v is large compared to other scales in the problem; we therefore neglect the last term in the equation, and get a 1-dimensional diffusion equation: $\frac{\partial T}{\partial t} = \kappa\frac{\partial^2 T}{\partial^2 z}$ with the initial conditions $T(t=0) = T_1 \cdot \Theta(-z)$.

The solution for $z \leq 0$ is given by the error function erf

$$T(x', z, t) = T_1 \cdot \mathrm{erf}\left(\frac{z}{2\sqrt{\kappa t}}\right)$$

Due to the large velocity, the temperature dependence on the horizontal direction is negligible, and the height at time t (i.e. of sea floor of age t) can be calculated by integrating the thermal expansion over z:

$$h(t) = h_0 + \alpha_{\mathrm{eff}} \int_0^{\infty} [T(z) - T_1] dz = h_0 - \frac{2}{\sqrt{\pi}} \alpha_{\mathrm{eff}} T_1 \sqrt{\kappa t}$$

where α_{eff} is the effective volumetric thermal expansion coefficient, and h_0 is the mid-ocean ridge height (compared to some reference).

Note that the assumption the v is relatively large is equivalently to the assumption that the thermal diffusivity κ is small compared to L^2/T, where L is the acean width (from mid-ocean ridges to continental shelf) and T is its age.

The effective thermal expansion coefficient

α_{eff} is different from the usual thermal expansion coefficient α

$$\alpha_{\mathrm{eff}} = \alpha \cdot \frac{\rho}{\rho - \rho_w}$$

where $\rho \sim 3.3 g/cm^3$ is the rock density and $\rho_0 = 1 g/cm^3$ is the density of water.

By substituting the parameters by their rough estimates: $\kappa \sim 8 \cdot 10^{-7}$ m^2/s, $\alpha \sim 4 \cdot 10^{-5}$ $°C^{-1}$ and T_1 ~1220 °C (for the Atlantic and Indian oceans) or ~1120 °C (for the eastern Pacific), we have:

$$h(t) \sim h_0 - 350\sqrt{t}$$

for the eastern Pacific Ocean, and:

$$h(t) \sim h_0 - 390\sqrt{t}$$

for the Atlantic and Indian Ocean, where the height is in meters and time is in millions of years. To get the dependence on x, one must substitute t = x/v ~ Tx/L, where L is the distance between the ridge to the continental shelf (roughly half the ocean width), and T is the ocean age.

Technical Terms

The acronym "mbsf" meaning "meters below the seafloor" is a convention used for depths below the seafloor.

BENTHOS

Benthos is the community of organisms that live on, in, or near the seabed, also known as the benthic zone. This community lives in or near marine sedimentary environments, from tidal pools along the foreshore, out to the continental shelf, and then down to the abyssal depths.

Many organisms adapted to deep-water pressure cannot survive in the upper parts of the water column. The pressure difference can be very significant (approximately one atmosphere for each 10 metres of water depth). Because light is absorbed before it can reach deep ocean-water, the energy source for deep benthic ecosystems is often organic matter from higher up in the water column that drifts down to the depths. This dead and decaying matter sustains the benthic food chain; most organisms in the benthic zone are scavengers or detritivores.

The term benthos, coined by Haeckel in 1891, comes from the Greek noun βένθος "depth of the sea". Benthos is also used in freshwater biology to refer to organisms at the bottom of freshwater bodies of water, such as lakes, rivers, and streams. There is also a redundant synonym, benthon.

Seaweed and two chitons in a tide pool.

Food Sources

Effect of eutrophication on marine benthic life

The main food sources for the benthos are algae and organic runoff from land. The depth of water, temperature and salinity, and type of local substrate all affect what benthos is present. In coastal waters and other places where light reaches the bottom, benthic photosynthesizing diatoms can proliferate. Filter feeders, such as sponges and bivalves, dominate hard, sandy bottoms. Deposit feeders, such as polychaetes, populate softer bottoms. Fish, such as dragonets, as well as sea stars, snails, cephalopods, and crustaceans are important predators and scavengers.

Benthic organisms, such as sea stars, oysters, clams, sea cucumbers, brittle stars and sea anemones, play an important role as a food source for fish, such as the California sheephead, and humans.

By Size

Macrobenthos

They are easily visible to the naked eye with the lower range of body size at 0.5 mm but usually larger than 3 mm. In the coastal water ecosystem, they include several species of organisms from different taxa including Porifera, Annelids, Coelenterates, Mollusks, Crustaceans, Arthropods etc.

Seagrass growing off the coast

Echinoderms

A sea squirt being used as a substrate for a nudibranch's spiral egg

Microphotograph of typical macrobenthic animals, (from top to bottom) including amphipods, a polychaete worm, a snail, and a chironomous midge larvae

MEIOBENTHOS

Meiobenthos, also called meiofauna, are small benthic invertebrates that live in both marine and fresh water environments. The term meiofauna loosely defines a group of organisms by their size, larger than microfauna but smaller than macrofauna, rather than a taxonomic grouping. In practice, that is organisms that can pass through a 1 mm mesh but will be retained by a 45 μm mesh, but the exact dimensions will vary from researcher to researcher. Whether an organism will pass through a 1 mm mesh will also depend upon whether it is alive or dead at the time of sorting.

The term meiobenthos was first coined in 1942 by Mare, but organisms that would fit into the meiofauna category have been studied since the 18th century. A comprehensive text on meiofauna is Introduction to the study of meiofauna by Higgins and Thiel (1988).

Collecting the Meiobenthos

Meiofauna are most commonly encountered in sedimentary environments in both marine and fresh water environments, from the littoral to the deep-sea. They can also be found on hard substrates living on algae, the phytal environment, and sessile invertebrates (barnacles, mussel beds, etc.).

Sampling Methodologies

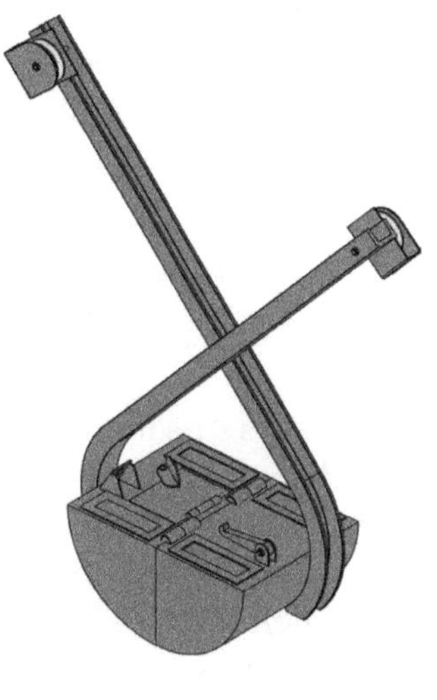

Van Veen grab

Sampling the meiobenthos is clearly dependent upon the environment and whether quantitative or qualititative samples are required. In the sedimentary environment the methodology used also depends on the physical morphology of the sediment. For qualititative sampling within the littoral zone, for both coarse and fine sediment, a bucket and spade will work. In the sub-littoral and deep water some form of grab (like the Van Veen Grab Sampler) is required, although a fine mesh (about 0.25 mm or less) would work also.

For the quantitative sampling of sedimentary environments at all depths a wide variety of samplers have been devised. The simplest is a plastic syringe with the end cut off to form a piston corer which can be deployed in the littoral zone, or in the sub-littoral using SCUBA gear. Generally the deeper the water the more complicated the sampling process becomes. For sampling the meiofauna on hard substrates, phytal and epizooic environments, the only practical methodology is to cut or scrape off a known area of substrate and place it in a plastic bag.

Extraction Methodologies

There are a wide variety of methods for extracting meiofauna from the samples of their habitat depending upon whether live or fixed specimens are required. For extracting live meiofauna one has to contend with the large number of species that cling or attach themselves to the substrate when disturbed. In order to get the meiofauna to release their grip there are three methodologies available.

The first, and simplest, is osmotic shock, this is achieved by submerging the sample in fresh water (clearly this will only work for marine samples) for a few seconds. This will cause the organisms to release after which they can then be shaken free from the substrate and filtered out through a 45 μm mesh and immediately returned to fresh filtered seawater. Many organisms will come through this process unharmed as long as the osmotic shock does not last too long.

The second methodology is the use of an anaesthetic. The preferred solution for meiobenthologists is isotonic magnesium chloride (7.5g $MgCl_2 \cdot 6H_2O$ in 100 ml of distilled water). The sample is immersed in the isotonic solution and left for a period of 15 min, after which the meiofauna are shaken free of the substrate and again filtered out through a 45 μm mesh and immediately returned to fresh filtered seawater.

The third methodology is Uhlig›s seawater ice technique. This relies on the organisms moving ahead of a front of ice cold seawater moving down through the sample ultimately forcing them out of the sediment. It is most effective on samples from temperate and tropical regions.

For major studies where large numbers of samples are collected concurrently, samples are normally fixed using 10% formalin solution and the meiofauna extracted at a later date. There are two main extraction methodologies. The first, decantation, works best with coarse sediments. Samples are shaken in an excess of water, the sediment is briefly allowed to settle and the meiofauna filtered off. The second methodology, the floatation technique, works best with finer sediments where the mass of the sediment particles is close to that of the meiofauna. The best solution to use is the colloidal silica, Ludox. The sample is stirred into the Ludox solution and left to settle for 40 min, after which the meiofauna are filtered out. In fine sediments, extraction efficiency is improved with centrifugation. With both methodologies repeated extractions should be made (at least three) with each sample to ensure that at least 95% of the fauna is extracted.

Meiofaunal Taxa

Based on the scheme of Nielsen (2001). Meiofaunal taxa appear in bold text.

- Protozoa
- Foraminifera
- Ciliophora
- Animalia—Eumetazoa
- Cnidaria
- Triploblastica—Bilateria
- Protostomia—Spiralia
- Schizocelia
- Sipuncula
- Articulata
- Mollusca
- Euarticulata
- Annelida
- Oligochaeta
- Polychaeta
- Panarthropoda
- Onychophora
- Arthropoda
- Tardigrada
- Amphipoda
- Cladocera
- Cumacea
- Harpacticoida
- Isopoda
- Ostracoda
- Mystacocarida
- Syncarida
- Tanaidacea
- Thermosbaenacea
- Halacarida
- Pycnogonida
- Palpigradida
- Insecta
- Bryozoa
- Entoprocta
- Ectoprocta

- Paranchymia
- Platyhelminthes (Turbellaria)
- Nemertea
- Gnathifera
- Rotifera
- Gnathostomulida
- Chaetognatha
- Cycloneuralia
- Gastrotricha
- Introverta
- Nematoda
- Nematophora
- Cephalorhyncha
- Priapulida
- Kinorhyncha
- Loricifera
- Deuterostomia
- Phoronida
- Brachiopoda
- Neorenalia
- Pterobranchia
- Echinodermata
- Cyrtotreta
- Entropneusta
- Chordata
- Urochordata
- Notochordata

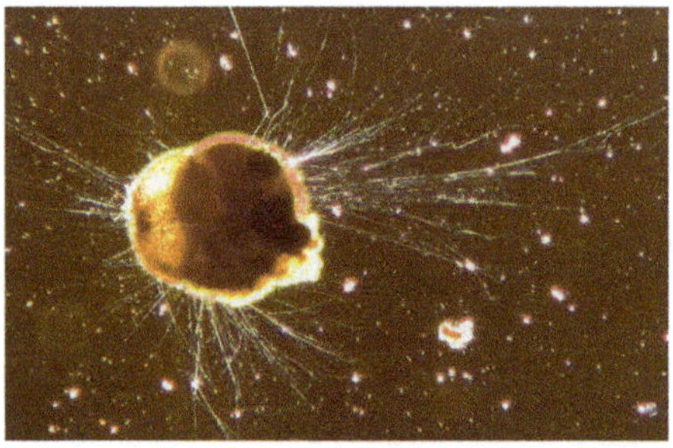

Live foraminifera Ammonia tepida (Rotaliida)

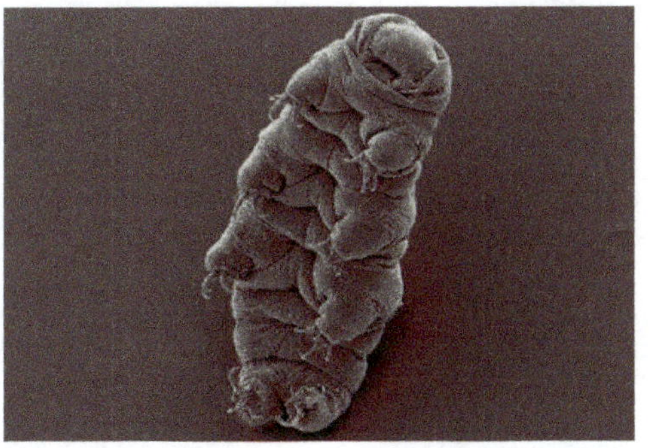

Water bear Hypsibius dujardini

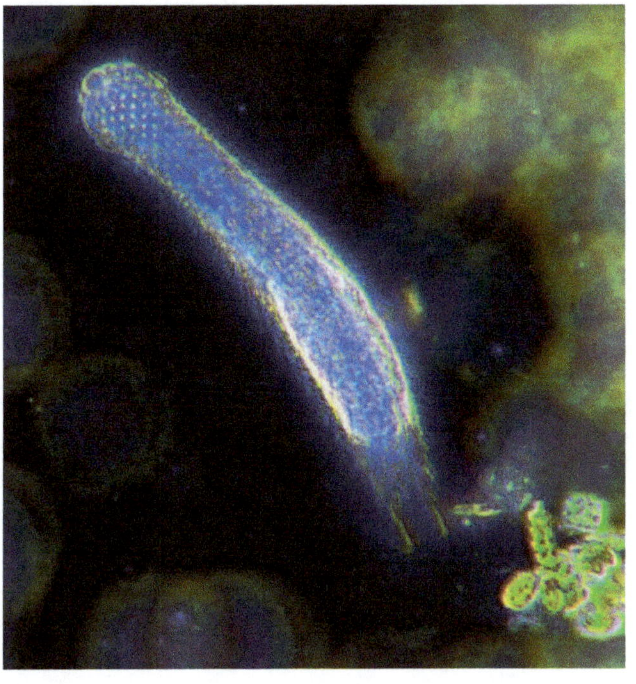

Gastrotrich

MICROBENTHOS

Microbenthos comprises microscopic benthic organisms that are less than 0.1 mm in size. Some examples are bacteria, diatoms, ciliates, amoeba, flagellates.

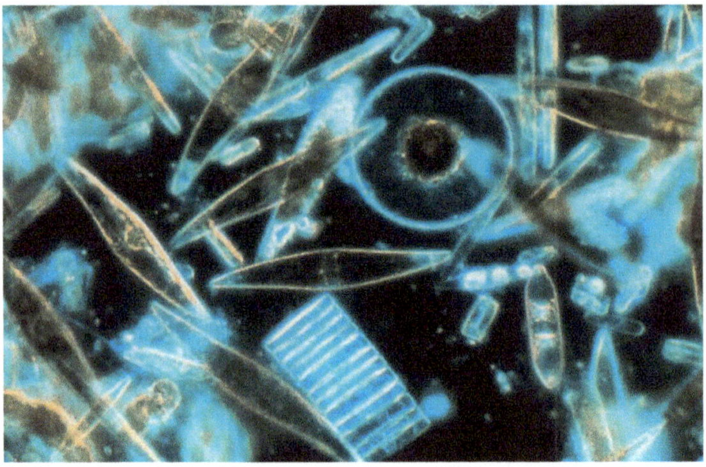

Marine diatoms

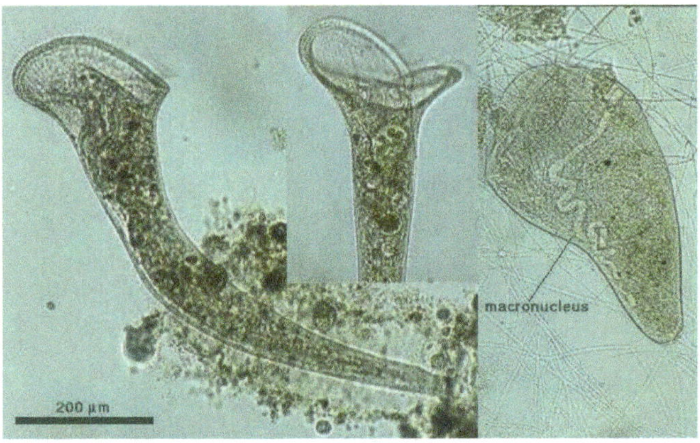

Ciliate stentor roeseli

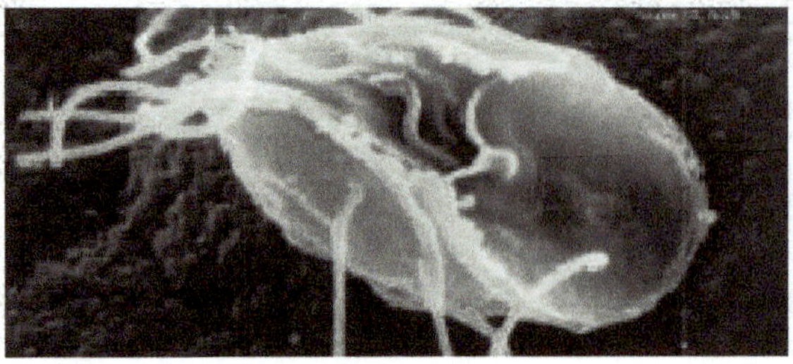

Flagellate

Benthos

Benthos is the community of organisms which live on, in, or near the seabed, the area known as the benthic zone. This community lives in or near marine sedimentary environments, from tidal pools along the foreshore, out to the continental shelf, and then down to the abyssal depths. The benthic zone is the ecological region on, in and immediately above the seabed, including the sediment surface and some sub-surface layers. Benthos generally live in close relationshipss with the substrate bottom, and many such organisms are permanently attached to the bottom. The superficial layer of the soil lining the given body of water, the benthic boundary layer, is an integral part of the benthic zone, and greatly influences the biological activity which takes place there. Examples of contact soil layers include sand bottoms, rocky outcrops, coral, and bay mud.

SEABED FEATURES

Each area of the seabed has typical features such as common soil composition, typical topography, salinity of water layers above it, marine life, magnetic direction of rocks,and sedimenting.

Seabed topography is flat where sedimenting is heavy and covers the tectonic features. Sedimenting comes from various sources:

- Land erosion sediments, brought mainly by rivers,
- New "young rock" – New magma of basalt composition, from the mid-ocean ridge.
- Underwater volcanic ash spreading, especially from hydrothermal vents.
- Microorganism activity.
- Sea currents eroding the seabed itself,
- Marine life: corals, fish, algae, crabs, marine plants and other biological created sediment.

Where sedimenting is avoided, such as in the Atlantic ocean especially in the northern and eastern Atlantic, the original tectonic activity can be clearly seen as straight line "cracks" or "vents" thousands of kilometers long.

Recently there has been the discovery of abundant marine life in the deep sea, especially around hydrothermal vents. Large deep sea communities of marine life have been discovered around black and white smokers – hydrothermal vents emitting typical chemicals toxic to humans and most of the vertebrates. This marine life receives its energy from both the extreme temperature difference (typically a drop of 150 degrees) and from chemosynthesis by bacteria.

Brine pools are another seabed feature, usually connected to cold seeps.

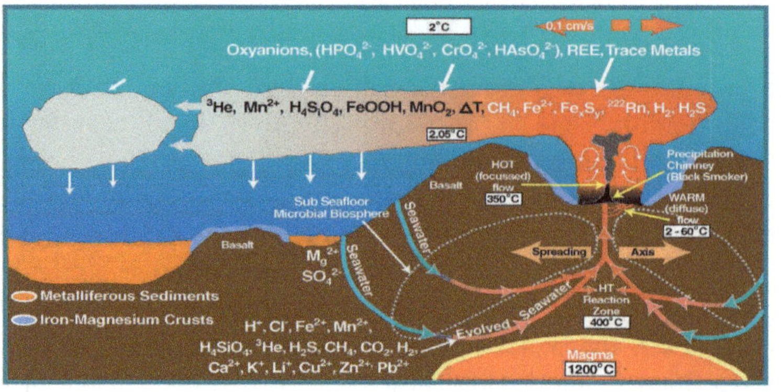

Oceanic ridge with deep sea vent

Layers of the pelagic zone

HISTORY OF EXPLORATION

The seabed has been explored by submersibles such as Alvin and, to some extent, scuba divers with special equipment. The process that continually adds new material to the ocean floor is seafloor spreading and the continental slope. In recent years satellite images show a very clear mapping of the seabed, and are used extensively in the study and exploration of the ocean floor.

In Art and Culture

Some children's play songs include elements such as "There's a hole at the bottom of the sea", or "A sailor went to sea... but all that he could see was the bottom of the deep blue sea".

On and under the seabed are archaeological sites of historic interest, such as shipwrecks and sunken towns. This underwater cultural heritage is protected by the UNESCO Convention on the Protection of the Underwater Cultural Heritage. The convention aims at preventing looting and the destruction or loss of historic and cultural information by providing an international legal framework.

REFERENCES

1. Flood, Roger D.; Piper, D.J.W., Klaus, A., and Peterson, L.C. (Eds.) (1997). "Preface: Depth Below Seafloor Conventions". Proceedings of the Ocean Drilling Program, Scientific Results. 155: 3. doi:10.2973/odp.proc.sr.155.200.1997.
2. Parkes, R. John; Henrik Sass (2007). Sulphate-reducing bacteria environmental and engineered systems. Edited by Larry L. Barton University of New Mexico. Sulphate-reducing bacteria environmental and engineered systems. Cambridge University Press. pp. 329–358. doi:10.1017/CBO9780511541490.012.
3. Safeguarding the Underwater Cultural Heritage UNESCO. Retrieved 12 September 2012.
4. Hess, H. H. (November 1962). "History of Ocean Basins" (PDF). In A. E. J. Engel; Harold L. James; B. F. Leonard. Petrologic studies: a volume to honor A. F. Buddington. Boulder, CO: Geological Society of America. pp. 599–620. Retrieved 8 September 2010.
5. Elsasser, Walter M. (1971). "Sea-Floor Spreading as Thermal Convection". Journal of Geophysical Research. 76: 1101. Bibcode:1971JGR....76.1101E. doi:10.1029/JB076i005p01101.
6. Makris, J.; Ginzburg, A. (1987-09-15). "Sedimentary basins within the Dead Sea and other rift zones The Afar Depression: transition between continental rifting and sea-floor spreading". Tectonophysics. 141 (1): 199–214. doi:10.1016/0040-1951(87)90186-7.
7. Bastow, Ian D.; Keir, Derek; Daly, Eve (2011-06-01). "The Ethiopia Afar Geoscientific Lithospheric Experiment (EAGLE): Probing the transition from continental rifting to incipient seafloor spreading". Geological Society of America Special Papers. 478: 51–76. doi:10.1130/2011.2478(04). ISSN 0072-1077.
8. Bhagwat, S.B. (2009). Foundation of Geology Vol 1. Global Vision Publishing House. p. 83. ISBN 9788182202764.
9. Vine, F. J.; Matthews, D. H. (1963). "Magnetic Anomalies Over Oceanic Ridges". Nature. 199: 974–949. doi:10.1038/199947a0.
10. Patriat, Philippe; Achache, José (1984). "India–Eurasia collision chronology has implications for crustal shortening and driving mechanism of plates". Nature. 311 (5987): 615. Bibcode:1984Natur.311..615P. doi:10.1038/311615a0.
11. Davis, E.E; Lister, C. R. B. (1974). "Fundamentals of Ridge Crest Topography". Earth and Planetary Science Letters. North-Holland Publishing Company. 21: 405–413. Bibcode:1974E&PSL..21..405D. doi:10.1016/0012-821X(74)90180-0.

CHAPTER 3

Ocean Zones

Chapter Contents

- **Introduction**
- **Description**
- **Deep Ocean Water**
- **Deep Sea**
- **Littoral Zone**
- **Mesopelagic Zone**
- **Oceanic Zone**
- **Pelagic Zone**
- **Photic Zone**
- **Surf Zone**
- **Swash**

There are four major oceanic zones where plants and animals live in the ocean. The four major zones are intertidal zone, neritic zone, open ocean zone and benthic zone. These zones contain the largest ecosystem on Earth.

INTRODUCTION

The benthic zone is the ecological region at the lowest level of a body of water such as an ocean or a lake, including the sediment surface and some sub-surface layers. Organisms living in this zone are called benthos, e.g. the benthic invertebrate community, including crustaceans and polychaetes. The organisms generally live in close relationship with the substrate bottom and many are permanently attached to the bottom. The superficial layer of the soil lining the given body of water, the benthic boundary layer, is an integral part of the benthic zone, as it greatly influences the biological activity that takes place there. Examples of contact soil layers include sand bottoms, rocky outcrops, coral, and bay mud.

DESCRIPTION

The benthic region of the ocean begins at the shore line (intertidal or eulittoral zone) and extends downward along the surface of the continental shelf out to sea. The continental shelf is a gently sloping benthic region that extends away from the land mass. At the continental shelf edge, usually about 200 meters deep, the gradient greatly increases and is known as the continental slope. The continental slope drops down to the deep sea floor. The deep-sea floor is called the abyssal plain and is usually about 4,000 meters deep. The ocean floor is not all flat but has submarine ridges and deep ocean trenches known as the hadal zone.

For comparison, the pelagic zone is the descriptive term for the ecological region above the benthos, including the water-column up to the surface. Depending on the water-body, the benthic zone may include areas that are only a few inches below water, such as a stream or shallow pond; at the other end of the spectrum, benthos of the deep ocean includes the bottom levels of the oceanic abyssal zone.

For information on animals that live in the deeper areas of the oceans see aphotic zone. Generally, these include life forms that tolerate cool temperatures and low oxygen levels, but this depends on the depth of the water.

Organisms

Benthos are the organisms that live in the benthic zone, and are different from those elsewhere in the water column. Many are adapted to live on the substrate (bottom). In their habitats they can be considered as dominant creatures, but they are often a source of prey for Carcharhinidae such as the lemon shark. Many organisms adapted to deep-water pressure cannot survive in the upper parts of the water column. The pressure difference can be very significant (approximately one atmosphere for each 10 meters of water depth).

Because light does not penetrate very deep into ocean-water, the energy source for the benthic ecosystem is often organic matter from higher up in the water column that drifts down to the depths. This dead and decaying matter sustains the benthic food chain; most organisms in the benthic zone are scavengers or detritivores. Some microorganisms use chemosynthesis to produce biomass.

Benthic organisms can be divided into two categories based on whether they make their home on the ocean floor or an inch or two into the ocean floor. Those living on the surface of the ocean floor are known as epifauna. Those who live burrowed into the ocean floor are known as infauna. Extremophiles, including piezophiles, which thrive in high pressures, may also live there.

Nutrient Flux

Sources of food for benthic communities can derive from the water column above these habitats in the form of aggregations of detritus, inorganic matter, and living organisms. These aggregations are commonly referred to as marine snow, and are important for the deposition of organic matter, and bacterial communities. The amount of material sinking to the ocean floor can average 307,000 aggregates per m^2 per day. This amount will vary on the depth of the benthos, and the degree of benthic-pelagic coupling. The benthos in a shallow region will have more available food than the benthos in the deep sea. Because of their reliance on it, microbes may become spatially dependent on detritus in the benthic zone. The microbes found in the benthic zone, specifically dinoflagellates and foraminifera, colonize quite rapidly on detritus matter while forming a symbiotic relationship with each other.

Habitats

Modern seafloor mapping technologies have revealed linkages between seafloor geomorphology and benthic habitats, in which suites of benthic communities are associated with specific geomorphic settings. Examples include cold-water coral communities associated with seamounts and submarine canyons, kelp forests associated with inner shelf rocky reefs and rockfish associated with rocky escarpments on continental slopes. In oceanic environments, benthic habitats can also be zoned by depth. From the shallowest to the deepest are: the epipelagic (less than 200 meters), the mesopelagic (200–1,000 metres), the bathyal (1,000–4,000 meters), the abyssal (4,000–6,000 meters) and the deepest, the hadal (below 6,000 meters).

The lower zones are in deep, pressurized areas of the ocean. Human impacts have occurred at all ocean depths, but are most significant on shallow continental shelf and slope habitats. Many benthic organisms have retained their historic evolutionary characteristics. Some organisms are significantly larger than their relatives living in shallower zones, largely because of higher oxygen concentration in deep water.

It is not easy to map or observe these organisms and their habitats, and most modern observations are made using remotely operated underwater vehicles (ROVs), and rarely submarines.

Ecological Research

Benthic macroinvertebrates have many important ecological functions, such as regulating the flow of materials and energy in river ecosystems through their food web linkages. Because of this correlation between flow of energy and nutrients, benthic macroinvertebrates have the ability to influence food resources on fish and other organisms in aquatic ecosystems. For example, the addition of a moderate amount of nutrients to a river over the course of several years resulted in increases in invertebrate richness, abundance, and biomass. These in turn resulted in increased food resources for native species of fish with insignificant alteration of the macroinvertebrate community structure and trophic pathways. The presence of macroinvertebrates such as Amphipoda also affect the dominance of certain types of algae in Benthic ecosystems as well. In addition, because benthic zones are influenced by the flow of dead organic material, there have been studies conducted on the relationship between stream and river water flows and the resulting effects on the benthic zone. Low flow events show a restriction in nutrient transport from benthic substrates to food webs, and caused a decrease in benthic macroinvertebrate biomass, which lead to the disappearance of food sources into the substrate.

Because the benthic system regulates energy in aquatic ecosystems, studies have been made of the mechanisms of the benthic zone in order to better understand the ecosystem. Benthic diatoms have been used by the European Union's Water Framework Directive (WFD) to establish ecological quality ratios that determined the ecological status of lakes in the UK. Beginning research is being made on benthic assemblages to see if they can be used as indicators of healthy aquatic ecosystems. Benthic assemblages in urbanized coastal regions are not functionally equivalent to benthic assemblages in untouched regions.

Ecologists are attempting to understand the relationship between heterogeneity and maintaining biodiversity in aquatic ecosystems. Benthic algae has been used as an inherently good subject for studying short term changes and community responses to heterogeneous conditions in streams. Understanding the potential mechanisms involving benthic periphyton and the effects on heterogeneity within a stream may provide a better understanding of the structure and function of stream ecosystems. Benthic gross primary production (GPP) may be important in maintaining biodiversity hotspots in littoral zones in large lake ecosystems. However, the relative contributions of benthic habitats within specific ecosystems are poorly explored and more research is needed.

DEEP OCEAN WATER

Deep ocean water (DOW) is the name for cold, salty water found deep below the surface of Earth›s oceans. Ocean water differs in temperature and salinity. Warm surface water is generally saltier than the cooler deep or polar waters; in polar regions, the upper layers of ocean water are cold and fresh. Deep ocean water makes up about 90% of the volume of the oceans. Deep ocean water has a very uniform temperature, around 3,8° °C (38 °F), and a salinity of about 3.5% or as oceanographers state as 35 ppt (parts per thousand).

In specialized locations such as the Natural Energy Laboratory of Hawaii NELHA ocean water is pumped to the surface from approximately 900 metres (3000 feet) deep for applications in research, commercial and pre-commercial activities. DOW is typically used to describe ocean water at sub-thermal depths sufficient to provide a measurable difference in water temperature.

When deep ocean water is brought to the surface, it can be used for a variety of things. Its most useful property is its temperature. At the surface of the Earth, most water and air is well above 3 °C. The difference in temperature is indicative of a difference in energy. Where there is an energy gradient, skillful application of science and engineering can harness that energy for productive use by humans. Assuming the source of deep ocean water is environmentally friendly and replenished by natural mechanisms, it forms a more innovative basis for cleaner energy than current fossil-fuel-derived energy.

The simplest use of cold water is for air conditioning: using the cold water itself to cool air saves the energy that would be used by the compressors for traditional refrigeration. Another use could be to replace expensive desalination plants. When cold water passes through a pipe surrounded by humid air, condensation results. The condensate is pure water, suitable for humans to drink or for crop irrigation. Via a technology called Ocean thermal energy conversion, the temperature difference can be turned into electricity.

Cold-Bed Agriculture

During condensation or Ocean thermal energy conservation operations, the water does not reach ambient temperature, because a certain temperature gradient is required to make these processes viable. The water leaving those operations is therefore still colder than the surroundings, and a further benefit can be extracted by passing this water through underground pipes, thereby cooling agricultural soil. This reduces evaporation, and even causes water to condense from the atmosphere. This allows agricultural production where crops would normally not be able to grow. This technique is sometimes referred to as "cold agriculture" or "cold-bed agriculture".

DEEP SEA

The deep sea or deep layer is the lowest layer in the ocean, existing below the thermocline and above the seabed, at a depth of 1000 fathoms (1800 m) or more. Little or no light penetrates this part of the ocean, and most of the organisms that live there rely for subsistence on falling organic matter produced in the photic zone. For this reason, scientists once assumed that life would be sparse in the deep ocean, but virtually every probe has revealed that, on the contrary, life is abundant in the deep ocean.

From the time of Pliny until the late nineteenth century...humans believed there was no life in the deep. It took a historic expedition in the ship *Challenger* between 1872 and 1876 to prove Pliny wrong; its deep-sea dredges and trawls brought up living things from all depths that could be reached. Yet even in the twentieth century scientists continued to imagine that life at great depth was insubstantial, or somehow inconsequential. The eternal dark, the almost inconceivable pressure, and the extreme cold that exist below one thousand meters were, they thought, so forbidding as to have all but extinguished life. The reverse is in fact true....(Below 200 meters) lies the largest habitat on earth.

In 1960, the Bathyscaphe *Trieste* descended to the bottom of the Mariana Trench near Guam, at 10,911 m (35,797 ft; 6.780 mi), the deepest known spot in any ocean. If Mount Everest (8,848 metres) were submerged there, its peak would be more than a mile beneath the surface. The *Trieste* was retired, and for a while the Japanese remote-operated vehicle (ROV) Kaikō was the only vessel capable of reaching this depth. It was lost at sea in 2003. In May and June 2009, the hybrid-ROV (HROV) *Nereus* returned to the Challenger Deep for a series of three dives to depths exceeding 10900 meters.

It has been suggested that more is known about the Moon than the deepest parts of the ocean. Little was known about the extent of life on the deep ocean floor until the discovery of thriving colonies of shrimps and other organisms around hydrothermal vents in the late 1970s. Before the discovery of the undersea vents, it had been accepted that almost all life on earth obtained its energy (one way or another) from the sun. The new discoveries revealed groups of creatures that obtained nutrients and energy directly from thermal sources and chemical reactions associated with changes to mineral deposits. These organisms thrive in completely lightless and anaerobic environments in highly saline water that may reach 300 °F (150 °C), drawing their sustenance from hydrogen sulfide, which is highly toxic to almost all terrestrial life. The revolutionary discovery that life can exist under these extreme conditions changed opinions about the chances of there being life elsewhere in the universe. Scientists now speculate that Europa, one of Jupiter›s moons, may be able to support life beneath its icy surface, where there is evidence of a global ocean of liquid water.

Environmental Characteristics

Light

Natural light does not penetrate the deep ocean, with the exception of the upper parts of the mesopelagic. Since photosynthesis is not possible, plants cannot live in this zone. Since plants are the primary producers of almost all of earth's ecosystems, life in this area of the ocean must depend on energy sources from elsewhere. Except for the areas close to the hydrothermal vents, this energy comes from organic material drifting down from the photic zone. The sinking organic material is composed of algal particulates, detritus, and other forms of biological waste, which is collectively referred to as marine snow.

Pressure

Because pressure in the ocean increases by about 1 atmosphere for every 10 meters of depth, the amount of pressure experienced by many marine organisms is extreme. Until recent years, the scientific community lacked detailed information about the effects of pressure on most deep sea organisms because the specimens encountered arrived at the surface dead or dying and weren't observable at the pressures at which they lived. With the advent of traps that incorporate a special pressure-maintaining chamber, undamaged larger metazoan animals have been retrieved from the deep sea in good condition.

Salinity

Salinity is remarkably constant throughout the deep sea, at about 35 parts per thousand. There are some minor differences in salinity, but none that is ecologically significant, except in the Mediterranean and Red Seas.

Temperature

The two areas of greatest and most rapid temperature change in the oceans are the transition zone between the surface waters and the deep waters, the thermocline, and the transition between the deep-sea floor and the hot water flows at the hydrothermal vents. Thermoclines vary in thickness from a few hundred meters to nearly a thousand meters. Below the thermocline, the water mass of the deep ocean is cold and far more homogeneous. Thermoclines are strongest in the tropics, where the temperature of the epipelagic zone is usually above 20 °C. From the base of the epipelagic, the temperature drops over several hundred meters to 5 or 6 °C at 1,000 meters. It continues to decrease to the bottom, but the rate is much slower. Below 3,000 to 4,000 m, the water is isothermal between 0 and 3 °C. The cold water stems from sinking heavy surface water in the polar regions.

At any given depth, the temperature is practically unvarying over long periods of time. There are no seasonal temperature changes, nor are there any annual changes. No other habitat on earth has such a constant temperature.

Hydrothermal vents are the direct contrast with constant temperature. In these systems, the temperature of the water as it emerges from the «black smoker» chimneys may be as high as 400 °C (it is kept from boiling by the high hydrostatic pressure) while within a few meters it may be back down to 2 - 4 °C.

Biology

Regions below the epipelagic are divided into further zones, beginning with the *mesopelagic* which spans from 200 to 1000 meters below sea level, where so little light penetrates that primary production becomes impossible. Below this zone the deep sea begins, consisting of the aphotic *bathypelagic*, *abyssopelagic* and *hadopelagic*. Food consists of falling organic matter known as 'marine snow' and carcasses derived from the productive zone above, and is scarce both in terms of spatial and temporal distribution.

Instead of relying on gas for their buoyancy, many deep-sea species have jelly-like flesh consisting mostly of glycosaminoglycans, which provides them with very low density. It is also common among deep water squid to combine the gelatinous tissue with a flotation chamber filled with a coelomic fluid made up of the metabolic waste product ammonium chloride, which is lighter than the surrounding water.

The midwater fish have special adaptations to cope with these conditions—they are small, usually being under 25 centimetres (10 in); they have slow metabolisms and unspecialized diets, preferring to sit and wait for food rather than waste energy searching for it. They have elongated bodies with weak, watery muscles and skeletal structures. They often have extendable, hinged jaws with recurved teeth. Because of the sparse distribution and lack of light, finding a partner with which to breed is difficult, and many organisms are hermaphroditic.

Because light is so scarce, fish often have larger than normal, tubular eyes with only rod cells. Their upward field of vision allows them to seek out the silhouette of possible prey. Prey fish however also have adaptations to cope with predation. These adaptations are mainly concerned with reduction of silhouettes, a form of camouflage. The two main methods by which this is achieved are reduction in the area of their shadow by lateral compression of the body, and counter illumination via bioluminescence. This is achieved by production of light from ventral photophores, which tend to produce such light intensity to render the underside of the fish of similar appearance to the background light. For more sensitive vision in low light, some fish have a retroreflector behind the retina. Flashlight fish have this plus photophores, which combination they use to detect eyeshine in other fish (see *tapetum lucidum*).

Organisms in the deep sea are almost entirely reliant upon sinking living and dead organic matter which falls at approximately 100 meters per day. In addition, only about 1-3% of the production from the surface reaches the sea bed mostly in the form of marine snow. Larger food falls, such as whale carcasses, also occur and studies have shown that these may happen more often than currently believed. There are many scavengers that feed primarily or entirely upon large food falls and the distance between whale carcasses is estimated to only be 8 kilometers. In addition, there are a number of filter feeders that feed upon organic particles using tentacles, such as *Freyella elegans* (see family Freyellidae). Marine bacteriophages play an important role in cycling nutrients in deep sea sediments. They are extremely abundant (between $5x10^{12}$ and $1x10^{13}$ phages per square meter) in sediments around the world.

Chemosynthesis

There are a number of species that do not primarily rely upon dissolved organic matter for their food and these are found at hydrothermal vents. One example is the symbiotic relationship between the tube worm *Riftia* and chemosynthetic bacteria. It is this chemosynthesis that supports the complex communities that can be found around hydrothermal vents. These complex communities are one of the few ecosystems on the planet that do not rely upon sunlight for their supply of energy.

Exploration

Describing the operation and use of an autonomous lander (RV Kaharoa) in deep-sea research; the fish seen are the abyssal grenadier (*Coryphaenoides armatus*).

The deep sea is an environment completely unfriendly to mankind; it represents one of the least explored areas on Earth. Pressures even in the mesopelagic become too great for traditional exploration methods, demanding alternative approaches for deep sea research. Baited camera stations, small manned submersibles and ROVs (remotely operated vehicles) are three methods utilized to explore the ocean's depths. Because of the difficulty and cost of exploring this zone, current knowledge is limited. Pressure increases at approximately one atmosphere for every 10 meters meaning that some areas of the deep sea can reach pressures of above 1,000 atmospheres. This not only makes great depths very difficult to reach without mechanical aids, but also provides a significant difficulty when attempting to study any organisms that may live in these areas as their cell chemistry will be adapted to such vast pressures.

LITTORAL ZONE

The littoral zone is the part of a sea, lake or river that is close to the shore. In coastal environments the littoral zone extends from the high water mark, which is rarely inundated, to shoreline areas that are permanently submerged. It always includes this intertidal zone and is often used to mean the same as the intertidal zone. However, the meaning of "littoral zone" can extend well beyond the intertidal zone.

There is no single definition. What is regarded as the full extent of the littoral zone, and the way the littoral zone is divided into subregions, varies in different contexts (lakes and rivers have their own definitions). The use of the term also varies from one part of the world to another, and between different disciplines. For example, military commanders speak of the littoral in ways that are quite different from marine biologists.

The adjacency of water gives a number of distinctive characteristics to littoral regions. The erosive power of water results in particular types of landforms, such as sand dunes, and estuaries. The natural movement of the littoral along the coast is called the littoral drift. Biologically, the ready availability of water enables a greater variety of plant and animal life, and particularly the formation of extensive wetlands. In addition, the additional local humidity due to evaporation usually creates a microclimate supporting unique types of organisms.

The word "littoral" is used both as a noun and an adjective. It derives from the Latin noun litus, litoris, meaning "shore". (The doubled 't' is a late medieval innovation and the word is sometimes seen in the more classical-looking spelling 'litoral'.)

In Oceanography and Marine Biology

The littoral zone of an ocean is the area close to the shore and extending out to the edge of the continental shelf.

The intertidal zone of a beach is also part of the littoral zone.

Estuaries are also in the littoral zone.

In oceanography and marine biology, the idea of the littoral zone is extended roughly to the edge of the continental shelf. Starting from the shoreline, the littoral zone begins at the spray region just above the high tide mark. From here, it moves to the intertidal region between the high and low water marks, and then out as far as the edge of the continental shelf. These three subregions are called, in order, the supralittoral zone, the eulittoral zone and the sublittoral zone.

Supralittoral Zone

The *supralittoral zone* (also called the *splash*, *spray* or *supratidal zone*) is the area above the spring high tide line that is regularly splashed, but not submerged by ocean water. Seawater penetrates these elevated areas only during storms with high tides. Organisms here must cope also with exposure to fresh water from rain, cold, heat and predation by land animals and seabirds. At the top of this area, patches of dark lichens can appear as crusts on rocks. Some types of periwinkles, Neritidae and detritus feeding Isopoda commonly inhabit the lower supralittoral.

Eulittoral zone

The *eulittoral zone* (also called the *midlittoral* or *mediolittoral zone*) is the *intertidal zone* also known as the *foreshore*. It extends from the spring high tide line, which is rarely inundated, to the spring low tide line, which is rarely not inundated. The wave action and turbulence of recurring tides shapes and reforms cliffs, gaps, and caves, offering a huge range of habitats for sedentary organisms. Protected rocky shorelines usually show a narrow almost homogenous eulittoral strip, often marked by the presence of barnacles. Exposed sites show a wider extension and are often divided into further zones. For more on this, see intertidal ecology.

Sublittoral Zone

The *sublittoral zone* starts immediately below the eulittoral zone. This zone is permanently covered with seawater and is approximately equivalent to the neritic zone.

In physical oceanography, the sublittoral zone refers to coastal regions with significant tidal flows and energy dissipation, including non-linear flows, internal waves, river outflows and oceanic fronts. In practice, this typically extends to the edge of the continental shelf, with depths around 200 meters.

In marine biology, the sublittoral refers to the areas where sunlight reaches the ocean floor, that is, where the water is never so deep as to take it out of the photic zone. This results in high primary production and makes the sublittoral zone the location of the majority of sea life. As in physical oceanography, this zone typically extends to the edge of the continental shelf. The benthic zone in the sublittoral is much more stable than in the intertidal zone; temperature, water pressure, and the amount of sunlight remain fairly constant. Sublittoral corals do not have to deal with as much change as intertidal corals. Corals can live in both zones, but they are more common in the sublittoral zone.

Within the sublittoral, marine biologists also identify the following:

- The **infralittoral zone** is the algal dominated zone to maybe five metres below the low water mark.
- The **circalittoral zone** is the region beyond the infralittoral, that is, below the algal zone and dominated by sessile animals such as oysters.

Shallower regions of the sublittoral zone, extending not far from the shore, are sometimes referred to as the **subtidal zone**.

In Freshwater Ecosystems

Shoreline of a lake with nearly unvegetated littoral zone

In freshwater situations, littoral zones occur on the edge of large lakes and rivers, often with extensive areas of wetland. Hence, they are sometimes referred to as fringing wetlands. Here, the effects of tides are minimal, so other definitions of "littoral" are used. For example, the Minnesota Department of Natural Resources defines littoral as that portion of the lake that is less than 15 feet in depth.

The littoral zone may form a narrow or broad fringing wetland, with extensive areas of aquatic plants sorted by their tolerance to different water depths. Typically, four zones are recognized, from higher to lower on the shore: wooded wetland, wet meadow, marsh and aquatic vegetation. The relative areas of these four types depends not only on the profile of the shoreline, but upon past water levels. The area of wet meadow is particularly dependent upon past water levels; in general, the area of wet meadows along lakes and rivers increases with natural water level fluctuations. Many of the animals in lakes and rivers are dependent upon the wetlands of littoral zones, since the rooted plants provide habitat and food. Hence, a large and productive littoral zone is considered an important characteristic of a healthy lake or river.

Littoral zones are at particular risk for two reasons. First, human settlement is often attracted to shorelines, and settlement often disrupts breeding habitats for littoral zone species. For example, many turtles are killed on roads when they leave the water to lay their eggs in upland sites. Fish can be negatively affected by docks and retaining walls which remove breeding habitat in shallow water. Some shoreline communities even deliberately try to remove wetlands since they may interfere with activities like swimming. Overall, the presence of human settlement has a demonstrated negative impact upon adjoining wetlands. An equally serious problem is the tendency to stabilize lake or river levels with dams. Dams removed the spring flood which carries nutrients into littoral zones, and reduces the natural fluctuation of water levels upon which many wetland plants and animals depend. Hence, over time, dams can reduce the area of wetland from a broad littoral zone to a narrow band of vegetation. Marshes and wet meadows are at particular risk.

Other Definitions

For the purposes of naval operations, the United States Navy divides the littoral zone in the ways shown on the diagram at the top of this article. The United States Army Corps of Engineers and Environmental Protection Agency have their own definitions, and these have legal implications.

MESOPELAGIC ZONE

The Mesopelagic (Greek μέσον, middle) (also known as the middle pelagic or twilight zone) is that part of the pelagic zone that extends from a depth of 200 to 1000 metres (~660 to 3300 feet) below the ocean surface. It lies between the photic epipelagic above and the aphotic bathypelagic below, where there is no light at all.

Physical Conditions

Although the temperature varies less at any one height than the epipelagic, the mesopelagic is the location of the thermocline; and in warmer regions of the world, the temperature varies from over 20 °C (68 °F) at the top to around 4 °C (39 °F) at the boundary with the bathyal zone. Water generally moves slowly in the mesopelagic with a residency time of about a century though a variety of animals move vertically through the zone on a daily basis and various debris sink down in relatively short time frames.

Ecology

Although some light penetrates the mesopelagic zone, it is insufficient for photosynthesis. The general types of life forms found are daytime visiting herbivores, detritivores feeding on dead organisms and fecal pellets, and carnivores feeding on the former types.

Examples of animals in the mesopelagic zone are: bristlemouth, swordfish, squid, wolf eels, cuttlefish and other semi-deepsea creatures. The bristlemouth is the Earth›s most abundant vertebrate, numbering in the hundreds of trillions to quadrillions. The small amount of sunlight is sufficient for animals, such as the chain catshark, to be fluorescent.

A 2015 study found that the zone›s biomass is up to 10 times greater than earlier estimates, totaling 10 billion tons and representing some 90 percent of the planet's fish biomass, some 100 times that of the annual seafood catch and 200 times that of the world's 24 billion chickens.

As of 2015 its small commercial fishery targeted meal, oil and silage rather than human food.

Climate Change

The quantity of mesopelagic life is large enough to play a significant role in global carbon cycling. As the fish participate in the food chain, they acquire carbon rendered from carbon dioxide absorbed by surface waters. Upon death, they transport that carbon into the abyss. As of 2015 the revised estimates of mesopelagic life had yet to be incorporated into climate models.

OCEANIC ZONE

The oceanic zone begins in the area off shore where the water measures 200 meters (656 feet) deep or deeper.

It is the region of open sea beyond the edge of the continental shelf and includes 65% of the ocean's completely open water. The oceanic zone has a wide array of undersea terrain, including crevices that are often deeper than Mount Everest is tall, as well as deep-sea volcanoes and ocean basins. While it is often difficult for life to sustain itself in this type of environment, some species do thrive in the oceanic zone (Col).

Sub Zones

The oceanic zone is subdivided into the epipelagic, mesopelagic, and bathypelagic zones.

The epipelagic (euphotic) zone, also called the sunlit zone, receives enough sunlight to support photosynthesis. The temperatures in this zone range anywhere from 40 to −3 °C (104 to 27 °F)(NHPTV).

The mesopelagic (disphotic) zone, where only small amounts of light penetrate, lies below the epipelagic zone. This zone is often referred to as the Twilight Zone due to its scarce amount of light. Temperatures in the mesopelagic zone range from 5 to 4 °C (41 to 39 °F). The pressure is higher here, it can be up to 1,470 pounds per square inch (10,100,000 Pa) and increases with depth (NHPTV).

90% of the ocean lies in the bathypelagic (aphotic) zone into which no light penetrates. This is also called the midnight zone. Water pressure is very intense and the temperatures are near freezing (range 0 to 6 °C (32 to 43 °F))

Marine Life

Oceanographers have divided the ocean into zones based on how far light reaches. All of the light zones can be found in the oceanic zone. The epipelagic zone is the one closest to the surface and is the best lit. It extends to 200 meters and contains both phytoplankton and zooplankton that can support larger organisms like marine mammals and some types of fish. Past 200 meters, not enough light penetrates the water to support life, and no plant life exists (NHPTV).

There are creatures however, which thrive around hydrothermal vents, or geysers located on the ocean floor that expel super heated water that is rich in minerals.These organisms feed off of chemosynthetic bacteria, which use the super heated water and chemicals from the hydrothermal vents to create energy in place of photosynthesis. The existence of these bacteria allow creatures like squids, hatchet fish, octopuses, tube worms, giant clams, spider crabs and other organisms to survive (Knight).

Due to the total darkness in the zones past the epipelagic zone, many organisms that survive in the deep oceans do not have eyes, and other organisms make their own light with bioluminescence. Often the light is blue green in colour, because many marine organisms are sensitive to blue light. Two chemicals, luciferin and luciferase that react with one another to create a soft glow. The process by which bioluminescence is created is very similar to what happens when a glow stick is broken. Deep-sea organisms use bioluminescence for everything from luring prey to navigation (Knight).

Animals such as fish, whales, and sharks are found in the oceanic zone.

PELAGIC ZONE

Any water in a sea or lake that is neither close to the bottom nor near the shore can be said to be in the pelagic zone. The word «pelagic» is derived from Greek πέλαγος *(pélagos)*, meaning "open sea". The pelagic zone can be thought of in terms of an imaginary cylinder or water column that goes from the surface of the sea almost to the bottom. Conditions differ deeper in the water column such that as pressure increases with depth, the temperature drops and less light penetrates. Depending on the depth, the water column, rather like the Earth's atmosphere, may be divided into different layers.

The pelagic zone occupies 1,330 million km^3 (320 million mi^3) with a mean depth of 3.68 km (2.29 mi) and maximum depth of 11 km (6.8 mi). Fish that live in the pelagic zone are called pelagic fish. Pelagic life decreases with increasing depth. It is affected by light intensity, pressure, temperature, salinity, the supply of dissolved oxygen and nutrients, and the submarine topography, which is called bathymetry. In deep water, the pelagic zone is sometimes called the open-ocean zone and can be contrasted with water that is near the coast or on the continental shelf. In other contexts, coastal water not near the bottom is still said to be in the pelagic zone.

The pelagic zone can be contrasted with the benthic and demersal zones at the bottom of the sea. The benthic zone is the ecological region at the very bottom of the sea. It includes the sediment surface and some subsurface layers. Marine organisms living in this zone, such as clams and crabs, are called benthos. The demersal zone is just above the benthic zone. It can be significantly affected by the seabed and the life that lives there. Fish that live in the demersal zone are called demersal fish, which can be divided into benthic fish, which are denser than water so they can rest on the bottom, and benthopelagic fish, which swim in the water column just above the bottom. Demersal fish are also known as bottom feeders and groundfish.

Depth and Layers

Depending on how deep the sea is, the pelagic zone can extend over up to five horizontal layers in the ocean. From the top down, these are:

Epipelagic (Sunlight)

This is the illuminated zone at the surface of the sea where enough light is available for photosynthesis. Nearly all primary production in the ocean occurs here. Consequently, plants and animals are largely concentrated in this zone. Examples of organisms living in this zone are plankton, floating seaweed, jellyfish, tuna, many sharks and dolphins.

Mesopelagic (Twilight)

The name for this zone stems from Greek μέσον *(meson)*, meaning "middle". Although some light penetrates this second layer it is insufficient for photosynthesis. At about 500 m the water also becomes depleted of oxygen. Organisms survive this environment by having more efficient gills or by minimizing movement.

Examples of animals that live here are swordfish, squid, wolffish and some species of cuttlefish. Many organisms that live in this zone are bioluminescent. Some creatures living in the mesopelagic zone rise to the epipelagic zone at night to feed.

Bathypelagic (Midnight)

The name stems from Greek βαθύς *(bathýs)*, meaning "deep". At this depth, the ocean is pitch black, apart from occasional bioluminescent organisms, such as anglerfish. No living plant exists here. Most animals living here survive by consuming the detritus falling from the zones above, which is known as "marine snow», or, like the marine hatchetfish, by preying on other inhabitants of this zone. Other examples of this zone›s inhabitants are giant squid, smaller squids and the dumbo octopus. The giant squid is hunted here by deep-diving sperm whales.

Abyssopelagic (Lower Midnight)

The name is derived from Greek βυσσος *(ábyssos)*, meaning "bottomless" (a holdover from the times when the deep ocean, or abyss, was believed to be bottomless). Very few creatures live in the cold temperatures, high pressures and complete darkness of this depth. Among the species found in this zone are several species of squid; echinoderms including the basket star, swimming cucumber, and the sea pig; and marine arthropods including the sea spider. Many of the species living at these depths are transparent and eyeless because of the total lack of light in this zone.

Hadopelagic

The name is derived from the realm of Hades, the underworld in Greek mythology. This zone is mostly unknown, and very few species are known to live here (in the open areas). However, many organisms live in hydrothermal vents in this and other zones. Some define the hadopelagic as waters below 6,000 m (20,000 ft), whether in a trench or not. The bathypelagic, abyssopelagic, and hadopelagic zones are very similar in character, and some marine biologists combine them into a single zone or consider the latter two to be the same. The abyssal plain is covered with soft sludge composed of dead organisms from above.

Pelagic Ecosystem

The pelagic ecosystem is based on the phytoplankton which occupy the start of the foodchain. Phytoplankton manufacture their own food using a process of photosynthesis. Because they need sunlight, they inhabit the upper, sunlit epipelagic zone, which includes the coastal or neritic zone. Biodiversity diminishes markedly in the deeper zones below the epipelagic zone as dissolved oxygen diminishes, water pressure increases, temperatures become colder, food sources become scarce, and light diminishes and finally disappears.

The pelagic sooty tern spends months at a time flying at sea, returning to land only for breeding.

Pelagic Birds

Pelagic birds, also called **oceanic birds**, live on the open sea, rather than around waters adjacent to land or around inland waters. Pelagic birds feed on planktonic crustaceans, squid and forage fish. Examples are the Atlantic puffin, macaroni penguins, sooty terns, shearwaters, and procellariiforms such as the albatross, procellariids and petrels.

The term seabird includes birds which live around the sea adjacent to land, as well as pelagic birds.

Pelagic Fish

Pelagic fish live in the water column of coastal, ocean, and lake waters, but not on or near the bottom of the sea or the lake. They can be contrasted with demersal fish, which live on or near the bottom, and reef fish, which are associated with coral reefs.

These fish are often migratory forage fish, which feed on plankton, and the larger fish that follow and feed on the forage fish. Examples of migratory forage fish are herring, anchovies, capelin, and menhaden. Examples of larger pelagic fish which prey on the forage fish are billfish, tuna, and oceanic sharks.

Pelagic Invertebrates

Some examples of pelagic invertebrates include krill, copepods, jellyfish, decapod larvae, hyperiid amphipods, rotifers and cladocerans.

Thorson›s rule states that benthic marine invertebrates at low latitudes tend to produce large numbers of eggs developing to widely dispersing pelagic larvae, whereas at high latitudes such organisms tend to produce fewer and larger lecithotrophic (yolk-feeding) eggs and larger offspring.

Pelagic Reptiles

Pelamis platura, the pelagic sea snake, is the only one of the 65 species of marine snakes to spend its entire life in the pelagic zone. It bears live young at sea and is helpless on land. The species sometimes forms aggregations of thousands along slicks in surface waters. The pelagic sea snake is the world's most widely distributed snake species.

Many species of sea turtles spend the first years of their lives in the pelagic zone, moving closer to shore as they reach maturity.

PHOTIC ZONE

The photic zone, euphotic zone (Greek for "well lit": ε "well" + φς "light"), or sunlight zone is the depth of the water in a lake or ocean that is exposed to such intensity of sunlight which designates compensation point, i.e. the intensity of light at which the rate of carbon dioxide uptake, or equivalently, the rate of oxygen production, is equal to the rate of carbon dioxide production, equivalently to the rate of oxygen consumption, reducing thus the net carbon dioxide assimilation to zero.

It extends from the surface down to a depth where light intensity falls to one percent of that at the surface, called the euphotic depth. Accordingly, its thickness depends on the extent of light attenuation in the water column. Typical euphotic depths vary from only a few centimetres in highly turbid eutrophic lakes, to around 200 meters in the open ocean. It also varies with seasonal changes in turbidity.

Since the photic zone is where almost all of the photosynthesis occurs, the depth of the photic zone is generally proportional to the level of primary production that occurs in that area of the ocean. About 90% of all marine life lives in the photic zone. A small amount of primary production is generated deep in the abyssal zone around the hydrothermal vents which exist along some mid-oceanic ridges.

The zone which extends from the base of the euphotic zone to about 200 metres is sometimes called the disphotic zone. While there is some light, it is insufficient for photosynthesis, or at least insufficient for photosynthesis at a rate greater than respiration. The euphotic zone together with the disphotic zone coincides with the epipelagic zone. The bottommost zone, below the euphotic zone, is called the aphotic zone. Most deep ocean waters belong to this zone.

The transparency of the water, which determines the depth of the photic zone, is measured simply with a Secchi disk. It may also be measured with a photometer lowered into the water.

SURF ZONE

As ocean surface waves come closer to shore they break, forming the foamy, bubbly surface called surf. The region of breaking waves defines the surf zone. After breaking in the surf zone, the waves (now reduced in height) continue to move in, and they run up onto the sloping front of the beach, forming an uprush of water called swash. The water then runs back again as backswash. The nearshore zone where wave water comes onto the beach is the surf zone. The water in the surf zone, or breaker zone, is shallow, usually between 5 and 10 m (16 and 33 ft) deep; this causes the waves to be unstable.

Animal Life

The animals that often are found living in the surf zone are crabs, clams, and snails. Surf clams and mole crabs are two species that stand out as inhabitants of the surf zone. Both of these animals are very fast burrowers. The surf clam, also known as the variable coquina, is a filter feeder that uses its gills to filter microalgae, tiny zooplankton, and small particulates out of seawater. The mole crab is a suspension feeder that eats by capturing zooplankton with its antennae. All of these creatures burrow down into the sand to escape from being pulled into the ocean from the tides and waves. They also burrow themselves in the sand to protect themselves from predators. The surf zone is full of nutrients, oxygen, and sunlight which leaves the zone very productive with animal life.

Rip Currents

Rip current in the ocean. Rip currents are often very difficult to spot with one's bare eyes, take caution in any body of water

The surf zone can contain dangerous rip currents: strong local currents which flow offshore and pose a threat to swimmers. Rip-current outlooks use the following set of qualifications:

- *Low-risk rip currents*: Wind and/or wave conditions are not expected to support the development of rip currents; however, rip currents can sometimes occur, especially in the vicinity of jetties and piers. Know how to swim and heed the advice of lifeguards.

- *Moderate-risk rip currents*: Wind and/or wave conditions support stronger or more frequent rip currents. Only experienced surf swimmers should enter the water.
- *High-risk rip currents*: Wind and/or wave conditions support dangerous rip currents. Rip currents are life-threatening to anyone entering the surf.

SWASH

Swash, or forewash in geography,(the white part of the water) is known as a turbulent layer of water that washes up on the beach after an incoming wave has broken. The swash action can move beach materials up and down the beach, which results in the cross-shore sediment exchange. The time-scale of swash motion varies from seconds to minutes depending on the type of beach. Greater swash generally occurs on flatter beaches. The swash motion plays the primary role in the formation of morphological features and their changes in the swash zone. The swash action also plays an important role as one of the instantaneous processes in wider coastal morphodynamics.

There are two approaches that describe swash motions: (1) swash resulting from the collapse of high-frequency bores ($f>0.05$ Hz) on the beachface; and (2) swash characterised by standing, low-frequency ($f<0.05$ Hz) motions. Which type of swash motion prevails is dependent on the wave conditions and the beach morphology and this can be predicted by calculating the surf similarity parameter εb (Guza & Inman 1975)

$$\epsilon b = \frac{4\pi^2 Hb}{2gT^2 tan^2\beta},$$

Where Hb is the breaker height, g is gravity, T is the incident-wave period and tan β is the beach gradient. Values $\varepsilon b>20$ indicate dissipative conditions where swash is characterised by standing long-wave motion. Values $\varepsilon b<2.5$ indicate reflective conditions where swash is dominated by wave bores.

Swash

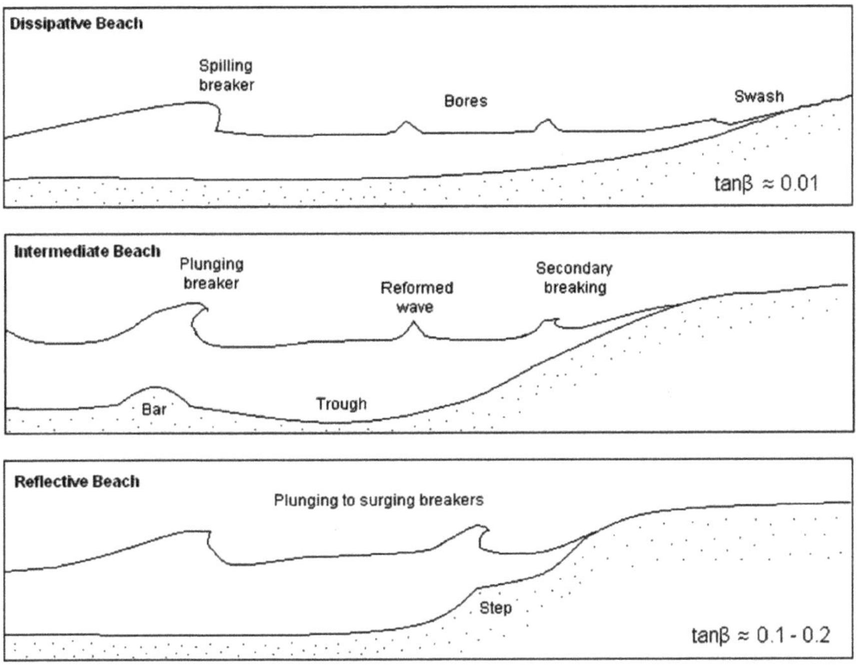

Beach classification by Wright and Short (1983) showing dissipative, intermediate, and reflective beaches.

Uprush and Backwash

Swash consists of two phases: **uprush** (onshore flow) and **backwash** (offshore flow). Generally uprush velocities are greater but of shorter duration compared to the backwash. Onshore velocities are at greatest at the start of the uprush and then decrease, whereas offshore velocities increase towards the end of the backwash. The direction of the uprush varies with the prevailing wind, whereas the backwash is always perpendicular to the coastline. This asymmetrical motion of swash can cause longshore drift as well as cross-shore sediment transport.

Swash Morphology

The swash zone is the upper part of the beach between backbeach and surf zone, where intense erosion occurs during storms. The swash zone is alternately wet and dry. Infiltration (hydrology) (above the water table) and exfiltration (below the water table) take place between the swash flow and the beach groundwater table. Beachface, berm, beach step and beach cusps are the typical morphological features associated with swash motion. Infiltration (hydrology) and sediment transport by swash motion are important factors that govern the gradient of the beachface.

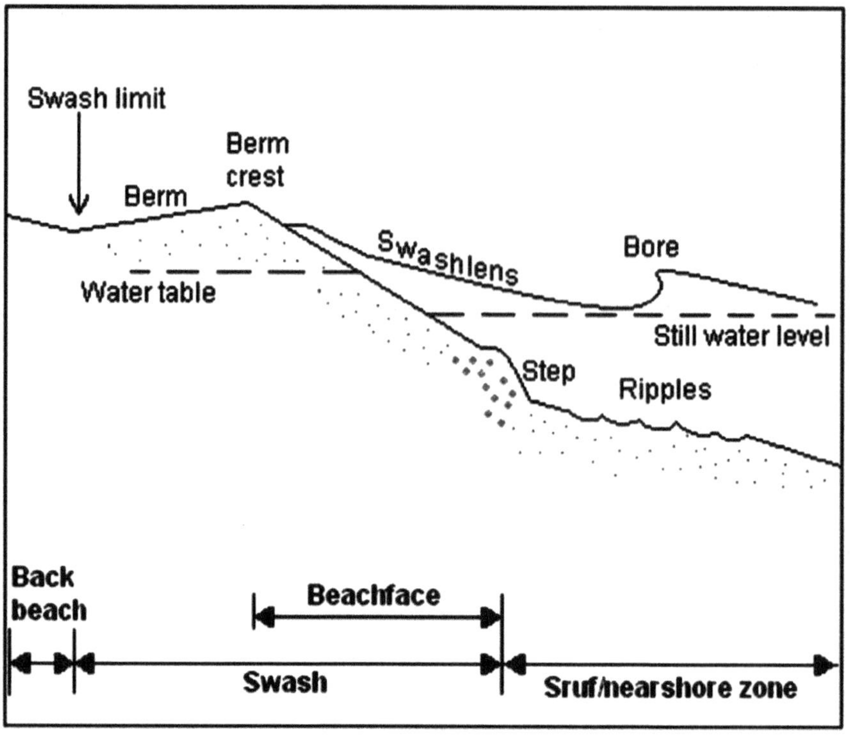

Swash zone and beachface morphology showing terminology and principal processes (Modified from Masselink & Hughes 2003)

Beachface

The beachface is the planar, relatively steep section of the beach profile that is subject to swash processes. The beachface extends from the berm to the low tide level. The beachface is in dynamic equilibrium with swash action when the amount of sediment transport by uprush and backwash are equal. If the beachface is flatter than the equilibrium gradient, more sediment is transported by the uprush to result in net onshore sediment transport. If the beachface is steeper than the equilibrium gradient, the sediment transport is dominated by the backwash and this results in net offshore sediment transport. The equilibrium beachface gradient is governed by a complex interrelationship of factors such as the sediment size, permeability, and fall velocity in the swash zone as well as the wave height and the wave period. The beachface cannot be considered in isolation from the surf zone to understand the morphological changes and equilibriums as they are strongly affected by the surf zone and shoaling wave processes as well as the swash zone processes.

Berm

The berm is the relatively planar part of the swash zone where the accumulation of sediment occurs at the landward farthest of swash motion. The berm protects the backbeach and coastal dunes from waves but erosion can occur under high energy conditions such as storms. The berm is more easily defined on gravel beaches and there can be multiple berms at different elevations. On sandy beaches in contrast, the gradient of backbeach, berm and beachface can be similar. The height of the berm is governed by the maximum elevation of sediment transport during the uprush. The berm height can be predicted using the equation by Takeda and Sunamura (1982)

$$Zberm = 0.125Hb^{5/8}(gT^2)^{3/8},$$

where Hb is the breaker height, *g* is gravity and T is the wave period.

Beach Step

The beach step is a submerged scarp at the base of the beachface. The beach steps generally comprise the coarsest material and the height can vary from several centimetres to over a metre. Beach steps form where the backwash interacts with the oncoming incident wave and generate vortex. Hughes and Cowell (1987) proposed the equation to predict the step height Zstep

$$Zstep = \sqrt{HbTws},$$

where 'ws' is the sediment fall velocity. Step height increases with increasing wave (breaker) height (Hb), wave period (T) and sediment size.

Beach Cusps

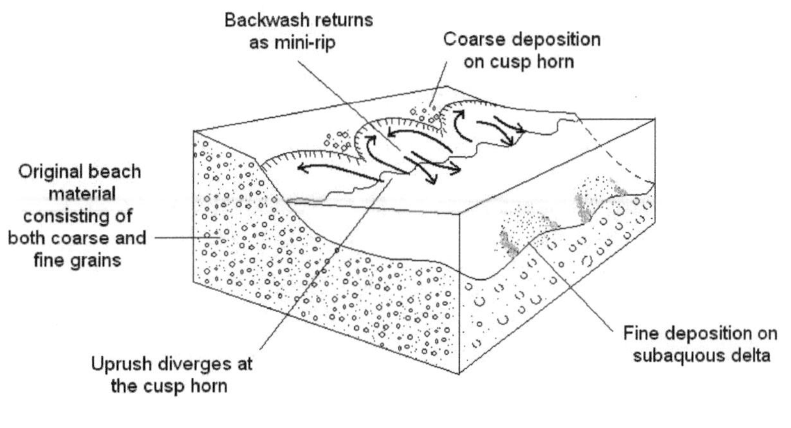

Beach cusp morphology. Uprush diverges at the cusp horns and backwash converges in the cusp embayments. (Modified from Masselink & Hughes 2003)

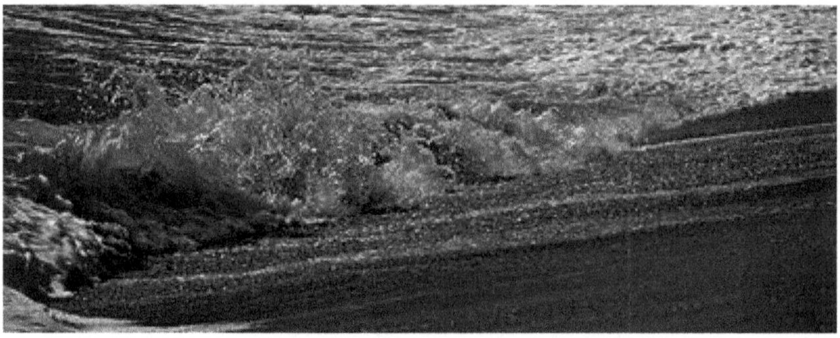

Backwash on a beach

The beach cusp is a crescent-shaped accumulation of sand or gravel surrounding a semicircular depression on a beach. They are formed by swash action and more common on gravel beaches than sand. The spacing of the cusps is related to the horizontal extent of the swash motion and can range from 10 cm to 50 m. Coarser sediments are found on the steep-gradient, seaward pointing 'cusp horns'. Currently there are two theories that provide an adequate explanation for the formation of the rhythmic beach cusps: standing edge waves and self-organization.

Standing Edge Wave Model

The standing edge wave theory, which was introduced by Guza and Inman (1975), suggests that swash is superimposed upon the motion of standing edge waves that travel alongshore. This produces a variation in swash height along the shore and consequently results in regular patterns of erosion. The cusp embayments form at the eroding points and cusp horns occur at the edge wave nodes. The beach cusp spacing can be predicted using the sub-harmonic edge wave model

$$\lambda = \frac{g}{\pi} T^2 tan\beta,$$

where T is incident wave period and $\tan\beta$ is beach gradient.

This model only explains the initial formation of the cusps but not the continuing growth of the cusps. The amplitude of the edge wave reduces as the cusps grow, hence it is a self-limiting process.

Self-Organization Model

The self-organization theory was introduced by Werner and Fink (1993) and it suggests that beach cusps form due to a combination of positive feedback that is operated by beach morphology and swash motion encouraging the topographic irregularity and negative feedback that discourages accretion or erosion on well-developed beach cusps. It is relatively recent that the computational resources and sediment transport formulations became available to show that the stable and rhythmic morphological features can be produced by such feedback systems. The beach cusp spacing, based on the self-organization model, is proportional to the horizontal extent of the swash motion S using the equation

$$\lambda = fS,$$

where the constant of proportionality f is *c.* 1.5.

Sediment Transport

Cross-Shore Sediment Transport

The cross-shore sediment exchange, between the subaerial and sub-aqueous zones of the beach, is primarily provided by the swash motion. The transport rates in the swash zone are much higher compared to the surf zone and suspended sediment concentrations can exceed 100 kg/m^3 close to the bed. The onshore and offshore sediment transport by swash thus plays a significant role in accretion and erosion of the beach.

There are fundamental differences in sediment transport between the uprush and backwash of the swash flow. The uprush, which is mainly dominated by bore turbulence, especially on steep beaches, generally suspend sediments to transport. Flow velocities, suspended sediment concentrations and suspended fluxes are at greatest at the start of the uprush when the turbulence is maximum. Then the turbulence dissipates towards the end of the onshore flow, settling the suspended sediment to the bed. In contrast, the backwash is dominated by the sheet flow and bedload sediment transport. The flow velocity increases towards the end of the backwash causing more bed-generated turbulence, which results in sediment transport near the bed. The direction of the net sediment transport (onshore or offshore) is largely governed by the beachface gradient.

Longshore Drift

Longshore drift by swash occurs either due to beach cusp morphology or due to oblique incoming waves causing strong alongshore swash motion. Under the influence of longshore drift, when there is no slack-water phase during backwash flows, sediments can remain suspended to result in offshore sediment transport. Beachface erosion by swash processes is not very common but erosion can occur where swash has a significant alongshore component.

Management

The swash zone is highly dynamic, accessible and susceptible to human activities. This zone can be very close to developed properties. It is said that at least 100 million people on the globe live within one meter of mean sea level. Understanding the swash zone processes and wise management is vital for coastal communities which can be affected by coastal hazards, such as erosion and storm surge. It is important to note that the swash zone processes cannot be considered in isolation as it is strongly linked with the surf zone processes. Many other factors, including human activities and climate change, can also influence the morphodynamics in the swash zone. Understanding the wider morphodynamics is essential in successful coastal management. Construction of sea walls has been a common tool to protect developed property, such as roads and buildings, from coastal erosion and recession. However, more often than not, protecting the property by building a seawall does not achieve the retention of the beach. Building an impermeable structure such as a seawall within the swash zone can interfere with the morphodynamics system in the swash zone. Building a seawall can raise the water table, increase wave reflection and intensify turbulence against the wall. This ultimately results in erosion of the adjacent beach or failure of the structure. Boulder ramparts (also known as revetments or riprap) and tetrapods are less reflective than impermeable sea walls, as waves are expected to break across the materials to produce swash and backwash that do not cause erosion. Rocky debris is sometimes placed in front of a sea wall in the attempt to reduce the wave impact, as well as to allow the eroded beach to recover.

Understanding the sediment transport system in the swash zone is also vital for beach nourishment projects. Swash plays a significant role in transportation and distribution of the sand that is added to the beach. There have been failures in the past due to inadequate understanding. Understanding and prediction of the sediment movements, both in the swash and surf zone, is vital for the nourishment project to succeed.

Example

The coastal management at Black Rock, on the north-east coast of Phillip Bay, Australia, provides a good example of a structural response to beach erosion which resulted in morphological changes in the swash zone. In the 1930s, a sea wall was built to protect the cliff from recession at Black Rock. This resulted in depletion of the beach in front of the sea wall, which was damaged by repeated storms in winter time. In 1969, the beach was nourished with approximately 5000m^3 of sand from inland in order to increase the volume of sand on the beach to protect the sea wall. This increased the sand volume by about 10%, however, the sand was carried away by northward drifting in autumn to leave the sea wall exposed to the impacts of winter storms again. The project had failed to take the seasonal patterns of longshore drift into account and had underestimated the amount of sand to nourish with, especially on the southern part of the beach.

Research

It is said that conduct of morphology research and field measurements in the swash zone is challenging since it is a shallow and aerated environment with rapid and unsteady swash flows. Despite the accessibility to the swash zone and the capability to take measurements with high resolution compared to the other parts of the nearshore zone, irregularity of the data has been an impediment for analysis as well as critical comparisons between theory and observation. Various and unique methods have been used for field measurements in the swash zone. For wave run-up measurements, for example, Guza and Thornton (1981, 1982) used an 80m long dual-resistance wire stretched across the beach profile and held about 3 cm above the sand by non-conducting supports. Holman and Sallenger (1985) conducted run-up investigation by taking videos of the swash to digitise the positions of the waterline over time. Many of the studies involved engineering structures, including seawalls, jetties and breakwaters, to establish design criteria that protect the structures from overtopping by extreme run-ups. Since the 1990s, swash hydrodynamics have been more actively investigated by coastal researchers, such as Hughes M.G., Masselink J. and Puleo J.A., contributing to the better understanding of the morphodynamics in the swash zone including turbulence, flow velocities, interaction with the beach groundwater table, and sediment transport. However, the gaps in understanding still remain in swash research including turbulence, sheet flow, bedload sediment transport and hydrodynamics on ultra-dissipative beaches.

Conclusion

Swash plays an important role as one of the instantaneous coastal processes and it is as important as the long-term processes such as sea level rise and geological processes in coastal morphodynamics. Swash zone is one of the most dynamic and rapidly changing environments on the coast and it is strongly linked with the surf zone processes. Understanding the swash mechanism is essential for the understanding of formation and changes of the swash zone morphology. More importantly, understanding of the swash zone processes is vital for society to manage coast wisely. There has been significant progress in the last two decades, however, gaps in understanding and knowledge in swash research still remain today.

REFERENCES

1. "What Are Benthos?". Baybenthos.versar.com. 2006-01-23. Retrieved 2013-11-24.
2. Angelo Mark P. Walag; Mae Oljae P. Canencia (2016). "Physico-chemical Parameters and Macrobenthic Invertebrates of the Intertidal Zone of Gusa, Cagayan de Oro City, Philippines". Advances in Environmental Sciences - International Journal of the Bioflux Society. 8 (1): 71–82. Retrieved 2015-12-08. (registration required)
3. Bright, Michael (2000). The private life of sharks: the truth behind the myth. Mechanicsburg, Pennsylvania: Stackpole Books. ISBN 0-8117-2875-7.
4. "Epifaunal - Definition and More from the Free Merriam-Webster Dictionary". Merriam-webster.com. 2012-08-31. Retrieved 2013-11-24.
5. "Infauna - Definition and More from the Free Merriam-Webster Dictionary". Merriam-webster.com. 2012-08-31. Retrieved 2013-11-24.
6. Alldredge, Alice; Silver, Mary W. (1988). "Characteristics, dynamics and significance of marine snow". Progress in Oceanography. 20: 41–82. doi:10.1016/0079-6611(88)90053-5.
7. Shanks, Alan; Trent, Jonathan D. (1980). "Marine snow: sinking rates and potential role in vertical flux". Deep-Sea Research. 27A (2): 137–143. doi:10.1016/0198-0149(80)90092-8.
8. Harris, P. T.; Baker, E. K. 2012. "GeoHab Atlas of seafloor geomorphic features and benthic habitats – synthesis and lessons learned", in: Harris, P. T.; Baker, E. K. (eds.), Seafloor Geomorphology as Benthic Habitat: GeoHab Atlas of seafloor

geomorphic features and benthic habitats. Elsevier, Amsterdam, pp. 871-890.

9. Harris, P. T.; Baker, E. K.; 2012. Seafloor Geomorphology as Benthic Habitat: GeoHab Atlas of seafloor geomorphic features and benthic habitats. Elsevier, Amsterdam, p. 947.
10. Harris, P. T., 2012. "Anthropogenic threats to benthic habitats", in: Harris, P. T.; Baker, E. K. (eds.), Seafloor Geomorphology as Benthic Habitat: GeoHab Atlas of seafloor geomorphic features and benthic habitats. Elsevier, Amsterdam, pp. 39-60.
11. Royal Belgian Institute of Natural Sciences, news item March 2005 ArchivedSeptember 28, 2011, at the Wayback Machine.
12. Minshall, Wayne; Shafii, Bahman; Price, William J.; Holderman, Charlie; Anders, Paul J.; Lester, Gary; Barrett, Pat. "Effects of nutrient replacement on benthic macroinvertebrates in an ultraoligotrophic reach of the Kootenai River, 2003–2010". Freshwater Science. doi:10.1086/677900. JSTOR 10.1086/677900.
13. Duffy, J. Emmett; Hay, Mark E. (2000-05-01). "Strong impacts of grazing amphipods on the organization of a benthic community". Ecological Monographs. 70 (2): 237–263. doi:10.1890/0012-9615(2000)070 0237:SIOGAO]2.0.CO;2. ISSN 0012-9615.
14. Rolls, Robert; Leigh, Catherine; Sheldon, Fran (2012). "Mechanistic effects of low-flow hydrology on riverine ecosystems: ecological principles and consequences of alteration". Freshwater Science. 31 (4): 1163–1186. doi:10.1899/12-002.1. JSTOR 10.1899/12-002.1.
15. Bennion, Helen; Kelly, Martyn G.; Juggins, Steve; Yallop, Marian L.; Burgess, Amy; Jamieson, Jane; Krokowski, Jan (2014). "Assessment of Ecological Status in UK lakes using benthic diatoms". Freshwater Science. 33 (2): 639–654. doi:10.1086/675447. JSTOR 10.1086/675447.
16. Lowe, Michael; Peterson, Mark S. (2014). "Effects of Coastal Urbanization on Salt-Marsh Faunal Assemblages in the Northern Gulf of Mexico". Marine and Coastal Fisheries: Dynamics, Management, and Ecosystem Science. 6: 89–107. doi:10.1080/19425120.2014.893467.
17. Wellnitz, Todd; Rader, Russell B. (2003). "Mechanisms influencing community composition and succession in mountain stream periphyton: interactions between scouring history, grazing, and irradiance". Journal of the North American Benthological Society. 22 (4): 528–541. doi:10.2307/1468350. JSTOR 1468350.
18. Althouse, Bryan; Higgins, Scott; Vander Zanden, Jake M. (2014). "Benthic and Planktonic primary production along a nutrient gradient in Green Bay, Lake Michigan, USA". Freshwater Science. 33 (2): 487–498. doi:10.1086/676314. JSTOR 10.1086/676314.
19. "Ocean Stratification". The Climate System. Columbia Univ. Retrieved 22 September 2015.
20. "The Hidden Meltdown of Greenland". Nasa Science/Science News. NASA. Retrieved 23 September 2015.

21. "Temperature of Ocean Water". UCAR. Retrieved 2012-09-05.
22. "Energinat - DOW Technologies - Cold Agriculture". www.energinat.com. Retrieved 14 July 2016.
23. "What is OTEC". OTEC News. the OTEC foundation. Retrieved 14 July 2016.
24. Costello, Mark John; Cheung, Alan; De Hauwere, Nathalie (2010). "Surface Area and the Seabed Area, Volume, Depth, Slope, and Topographic Variation for the World's Seas, Oceans, and Countries". Environmental Science & Technology. 44 (23): 8821–8. Bibcode:2010EnST...44.8821C. doi:10.1021/es1012752.
25. Charette, Matthew; Smith, Walter (2010). "The Volume of Earth's Ocean". Oceanography. 23 (2): 112–4. doi:10.5670/oceanog.2010.51. hdl:1912/3862.
26. Ocean's Depth and Volume Revealed OurAmazingPlanet, 19 May 2010.
27. http://www.marinebio.com/Oceans/open-ocean.asp
28. BirdLife International (2008). "Sterna fuscata". IUCN Red List of Threatened Species. Version 2008. International Union for Conservation of Nature. Retrieved 7 August 2009.
29. Walker P and Wood E (2005) The Open Ocean (volume in a series called Life in the sea), Infobase Publishing, ISBN 978-0-8160-5705-4.
30. Lal BV and Fortune K (2000) The Pacific Islands: An encyclopedia Page 8. University of Hawaii Press. ISBN 978-0-8248-2265-1.
31. Thorson, G (1957). "Bottom communities (sublittoral or shallow shelf)". In Hedgpeth, J.W. Treatise on Marine Ecology and Palaeoecology. Geological Society of America. pp. 461–534.
32. Mileikovsky, S. A. (1971). "Types of larval development in marine bottom invertebrates, their distribution and ecological significance: a re-evaluation". Marine Biology. 10 (3): 193–213. doi:10.1007/BF00352809.
33. Whittow, J.B. 2000, The penguin dictionary for physical geography. Penguin Books, London
34. Komar, P.D. 1998, Beach processes and sedimentation. Prentice-Hall, Englewood Cliffs
35. Wright, L.D. and Short, A.D. 1984, "Morphodynamic variability of surf zones and beaches: A synthesis". Marine Geology, 56, pp.93-118
36. Masselink, G. and Hughes M.G. 2003, Introduction to coastal processes and geomorphology, Hodder Arnold, London
37. Masselink, G. and Puleo, J.A. 2006, "Swash-zone morphodynamics". Continental Shelf Research, 26, pp.661-680
38. Masselink, G. and Hughes, M. 1998, "Field investigation of sediment transport in the swash zone". Continental Shelf Research 18, pp.1179-1199
39. Zhang, K., Douglas, B.C. and Leatherman, S.P. 2004, "Global warming and coastal erosion". Climatic Change, 64, pp.41-58
40. Rae, E. 2010, "Coastal Erosion and Deposition" in Encyclopedia of Geography. Sage publications, 21 March 2011, <http://www.sage-ereference.com/geography/

Article_n185.html>

41. Bird, E.C.F. 1996, Beach management. John Wiley & Sons, Chichester
42. Blenkinsopp, C.E., Turner, I.L., Masselink, G., Russell, P.E. 2011, “Swash zone sediment fluxes: Field observations”. Coastal Engineering, 58, pp.28-44

CHAPTER 4

Oceanic Landforms

Chapter Contents

- **Introduction**
- **Formation**
- **Coral Reef Fish**
- **Coastal Fish**

Coastal and oceanic landforms. ... Coastal landforms and Physical oceanography topographic landforms. Subcategories. A coral reef is a sea structure made out of living organisms that form a wall together. Stony corals are usually the main organism found in a reef, but other types of skeletal material also join together to form the thick walls. It is also "constructed" by the thousands of fish, sponges, and sea urchins grazing around the area. As these organisms move about, they actually help the reef settle and gain a more stable structure. Waves are also partly responsible for the wall's size and format, as they affect the geology of the formation.

INTRODUCTION

Coral reefs are diverse underwater ecosystems held together by calcium carbonate structures secreted by corals. Coral reefs are built by colonies of tiny animals found in marine waters that contain few nutrients. Most coral reefs are built from stony corals, which in turn consist of polyps that cluster in groups. The polyps belong to a group of animals known as Cnidaria, which also includes sea anemones and jellyfish. Unlike sea anemones, corals secrete hard carbonate exoskeletons which support and protect the coral polyps. Most reefs grow best in warm, shallow, clear, sunny and agitated waters.

Often called rainforests of the sea, shallow coral reefs form some of the most diverse ecosystems on Earth. They occupy less than 0.1% of the world›s ocean surface, about half the area of France, yet they provide a home for at least 25% of all marine species, including fish, mollusks, worms, crustaceans, echinoderms, sponges, tunicates and other cnidarians. Paradoxically, coral reefs flourish even though they are surrounded by ocean waters that provide few nutrients. They are most commonly found at shallow depths in tropical waters, but deep water and cold water corals also exist on smaller scales in other areas.

Coral reefs deliver ecosystem services to tourism, fisheries and shoreline protection. The annual global economic value of coral reefs is estimated between US$29.8-375 billion. However, coral reefs are fragile ecosystems, partly because they are very sensitive to water temperature. They are under threat from climate change, oceanic acidification, blast fishing, cyanide fishing for aquarium fish, sunscreen use, overuse of reef resources, and harmful land-use practices, including urban and agricultural runoff and water pollution, which can harm reefs by encouraging excess algal

FORMATION

Most of the coral reefs we can see today were formed after the last glacial period when melting ice caused the sea level to rise and flood the continental shelves. This means that most modern coral reefs are less than 10,000 years old. As communities established themselves on the shelves, the reefs grew upwards, pacing rising sea levels. Reefs that rose too slowly could become drowned reefs. They are covered by so much water that there was insufficient light. Coral reefs are found in the deep sea away from continental shelves, around oceanic islands and as atolls. The vast majority of these islands are volcanic in origin. The few exceptions have tectonic origins where plate movements have lifted the deep ocean floor on the surface.

In 1842 in his first monograph, *The Structure and Distribution of Coral Reefs*, Charles Darwin set out his theory of the formation of atoll reefs, an idea he conceived during the voyage of the *Beagle*. He theorized uplift and subsidence of the Earth's crust under the oceans formed the atolls.Darwin's theory sets out a sequence of three stages in atoll formation. It starts with a fringing reef forming around an extinct volcanic island as the island and ocean floor subsides. As the subsidence continues, the fringing reef becomes a barrier reef, and ultimately an atoll reef.

Darwin's theory starts with a volcanic island which becomes extinct

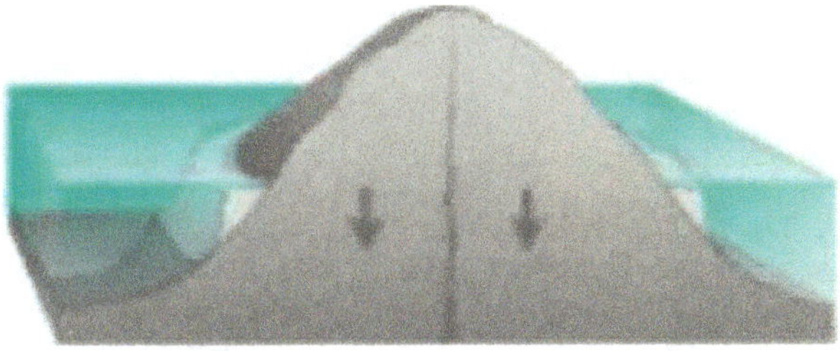

As the island and ocean floor subside, coral growth builds a fringing reef, often including a shallow lagoon between the land and the main reef.

As the subsidence continues, the fringing reef becomes a larger barrier reef further from the shore with a bigger and deeper lagoon inside.

Ultimately, the island sinks below the sea, and the barrier reef becomes an atoll enclosing an open lagoon.

Darwin predicted that underneath each lagoon would be a bed rock base, the remains of the original volcano. Subsequent drilling proved this correct. Darwin's theory followed from his understanding that coral polyps thrive in the clean seas of the tropics where the water is agitated, but can only live within a limited depth range, starting just below low tide. Where the level of the underlying earth allows, the corals grow around the coast to form what he called fringing reefs, and can eventually grow out from the shore to become a barrier reef.

A fringing reef can take ten thousand years to form, and an atoll can take up to 30 million years.

Where the bottom is rising, fringing reefs can grow around the coast, but coral raised above sea level dies and becomes white limestone. If the land subsides slowly, the fringing reefs keep pace by growing upwards on a base of older, dead coral, forming a barrier reef enclosing a lagoon between the reef and the land. A barrier reef can encircle an island, and once the island sinks below sea level a roughly circular atoll of growing coral continues to keep up with the sea level, forming a central lagoon. Barrier reefs and atolls do not usually form complete circles, but are broken in places by storms. Like sea level rise, a rapidly subsiding bottom can overwhelm coral growth, killing the coral polyps and the reef, due to what is called *coral drowning*. Corals that rely on zooxanthellae can *drown* when the water becomes too deep for their symbionts to adequately photosynthesize, due to decreased light exposure.

The two main variables determining the geomorphology, or shape, of coral reefs are the nature of the underlying substrate on which they rest, and the history of the change in sea level relative to that substrate.

The approximately 20,000-year-old Great Barrier Reef offers an example of how coral reefs formed on continental shelves. Sea level was then 120 m (390 ft) lower than in the 21st century. As sea level rose, the water and the corals encroached on what had been hills of the Australian coastal plain. By 13,000 years ago, sea level had risen to 60 m (200 ft) lower than at present, and many hills of the coastal plains had become continental islands. As the sea level rise continued, water topped most of the continental islands. The corals could then overgrow the hills, forming the present cays and reefs. Sea level on the Great Barrier Reef has not changed significantly in the last 6,000 years, and the age of the modern living reef structure is estimated to be between 6,000 and 8,000 years. Although the Great Barrier Reef formed along a continental shelf, and not around a volcanic island, Darwin›s principles apply. Development stopped at the barrier reef stage, since Australia is not about to submerge. It formed the world›s largest barrier reef, 300–1,000 m (980–3,280 ft) from shore, stretching for 2,000 km (1,200 mi).

Healthy tropical coral reefs grow horizontally from 1 to 3 cm (0.39 to 1.18 in) per year, and grow vertically anywhere from 1 to 25 cm (0.39 to 9.84 in) per year; however, they grow only at depths shallower than 150 m (490 ft) because of their need for sunlight, and cannot grow above sea level.

Materials

As the name implies, the bulk of coral reefs is made up of coral skeletons from mostly intact coral colonies. As other chemical elements present in corals become incorporated into the calcium carbonate deposits, aragonite is formed. However, shell fragments and the remains of calcareous algae such as the green-segmented genus *Halimeda* can add to the reef's ability to withstand damage from storms and other threats. Such mixtures are visible in structures such as Eniwetok Atoll.

Types

The three principal reef types are:

- **Fringing reef** – directly attached to a shore, or borders it with an intervening shallow channel or lagoon
- **Barrier reef** – reef separated from a mainland or island shore by a deep channel or lagoon
- **Atoll reef** – more or less circular or continuous barrier reef extends all the way around a lagoon without a central island

A small atoll in the Maldives

Inhabited cay in the Maldives

Other reef types or variants are:

- **Patch reef** – common, isolated, comparatively small reef outcrop, usually within a lagoon or embayment, often circular and surrounded by sand or seagrass
- **Apron reef** – short reef resembling a fringing reef, but more sloped; extending out and downward from a point or peninsular shore
- **Bank reef** – linear or semicircular shaped-outline, larger than a patch reef
- **Ribbon reef** – long, narrow, possibly winding reef, usually associated with an atoll lagoon
- **Table reef** – isolated reef, approaching an atoll type, but without a lagoon
- **Habili** – reef specific to the Red Sea; does not reach the surface near enough to cause visible surf; may be a hazard to ships (from the Arabic for "unborn")
- **Microatoll** – community of species of corals; vertical growth limited by average tidal height; growth morphologies offer a low-resolution record of patterns of sea level change; fossilized remains can be dated using radioactive carbon dating and have been used to reconstruct Holocene sea levels
- **Cays** – small, low-elevation, sandy islands formed on the surface of coral reefs from eroded material that piles up, forming an area above sea level; can be stabilized by plants to become habitable; occur in tropical environments throughout the Pacific, Atlantic and Indian Oceans (including the Caribbean and on the Great Barrier Reef and Belize Barrier Reef), where they provide habitable and agricultural land
- **Seamount** or **guyot** – formed when a coral reef on a volcanic island subsides; tops of seamounts are rounded and guyots are flat; flat tops of guyots, or *tablemounts*, are due to erosion by waves, winds, and atmospheric processes

Zones

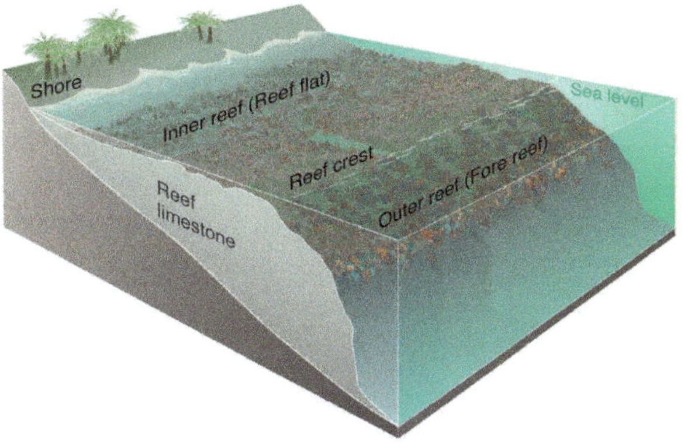

The three major zones of a coral reef: the fore reef, reef crest, and the back reef

Coral reef ecosystems contain distinct zones that represent different kinds of habitats. Usually, three major zones are recognized: the fore reef, reef crest, and the back reef (frequently referred to as the reef lagoon).

All three zones are physically and ecologically interconnected. Reef life and oceanic processes create opportunities for exchange of seawater, sediments, nutrients, and marine life among one another.

Thus, they are integrated components of the coral reef ecosystem, each playing a role in the support of the reefs' abundant and diverse fish assemblages.

Most coral reefs exist in shallow waters less than 50 m deep. Some inhabit tropical continental shelves where cool, nutrient rich upwelling does not occur, such as Great Barrier Reef. Others are found in the deep ocean surrounding islands or as atolls, such as in the Maldives. The reefs surrounding islands form when islands subside into the ocean, and atolls form when an island subsides below the surface of the sea.

Alternatively, Moyle and Cech distinguish six zones, though most reefs possess only some of the zones.

Water in the reef surface zone is often agitated. This diagram represents a reef on a continental shelf. The water waves at the left travel over the *off-reef floor* until they encounter the *reef slope* or *fore reef*. Then the waves pass over the shallow *reef crest*. When a wave enters shallow water it shoals, that is, it slows down and the wave height increases.

The reef surface is the shallowest part of the reef. It is subject to the surge and the rise and fall of tides. When waves pass over shallow areas, they shoal, as shown in the diagram at the right. This means the water is often agitated. These are the precise condition under which corals flourish. Shallowness means there is plenty of light for photosynthesis by the symbiotic zooxanthellae, and agitated water promotes the ability of coral to feed on plankton. However, other organisms must be able to withstand the robust conditions to flourish in this zone.

The off-reef floor is the shallow sea floor surrounding a reef. This zone occurs by reefs on continental shelves. Reefs around tropical islands and atolls drop abruptly to great depths, and do not have a floor. Usually sandy, the floor often supports seagrass meadows which are important foraging areas for reef fish.

The reef drop-off is, for its first 50 m, habitat for many reef fish who find shelter on the cliff face and plankton in the water nearby. The drop-off zone applies mainly to the reefs surrounding oceanic islands and atolls.

The reef face is the zone above the reef floor or the reef drop-off. This zone is often the most diverse area of the reef. Coral and calcareous algae growths provide complex habitats and areas which offer protection, such as cracks and crevices. Invertebrates and epiphytic algae provide much of the food for other organisms.

The reef flat is the sandy-bottomed flat, which can be behind the main reef, containing chunks of coral. This zone may border a lagoon and serve as a protective area, or it may lie between the reef and the shore, and in this case is a flat, rocky area. Fishes tend to prefer living in that flat, rocky area, compared to any other zone, when it is present.

The reef lagoon is an entirely enclosed region, which creates an area less affected by wave action that often contains small reef patches.

However, the topography of coral reefs is constantly changing. Each reef is made up of irregular patches of algae, sessile invertebrates, and bare rock and sand. The size, shape and relative abundance of these patches changes from year to year in response to the various factors that favor one type of patch over another. Growing coral, for example, produces constant change in the fine structure of reefs. On a larger scale, tropical storms may knock out large sections of reef and cause boulders on sandy areas to move.

Locations

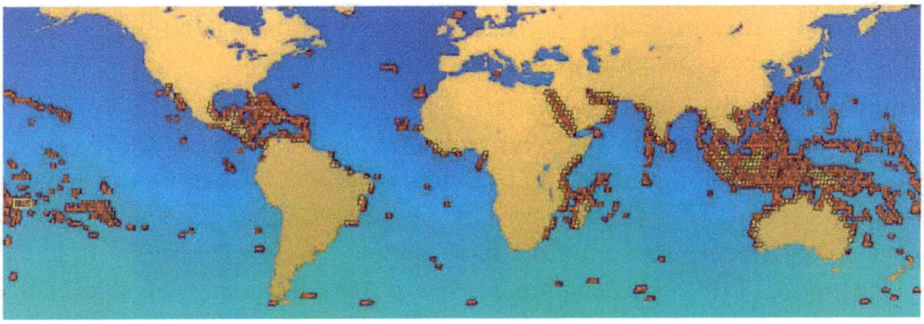

Locations of coral reefs

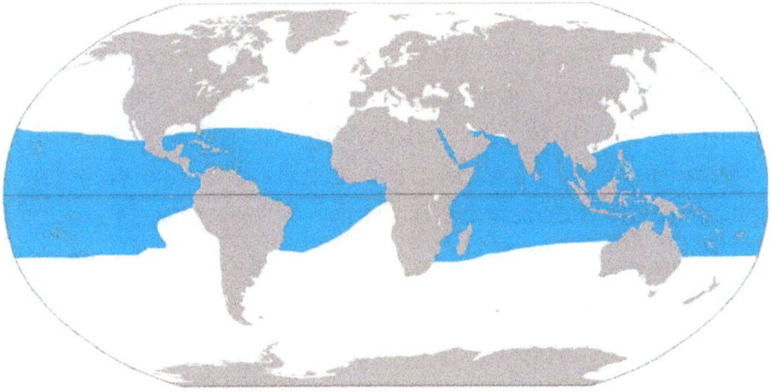

Boundary for 20 °C isotherms. Most corals live within this boundary. Note the cooler waters caused by upwelling on the southwest coast of Africa and off the coast of Peru.

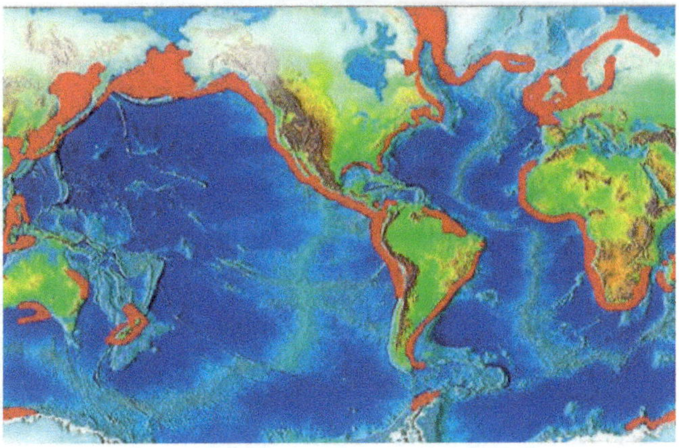

This map shows areas of upwelling in red. Coral reefs are not found in coastal areas where colder and nutrient-rich upwellings occur.

Coral reefs are estimated to cover 284,300 km^2 (109,800 sq mi), just under 0.1% of the oceans› surface area. The Indo-Pacific region (including the Red Sea, Indian Ocean, Southeast Asia and the Pacific) account for 91.9% of this total. Southeast Asia accounts for 32.3% of that figure, while the Pacific including Australia accounts for 40.8%. Atlantic and Caribbean coral reefs account for 7.6%.

Although corals exist both in temperate and tropical waters, shallow-water reefs form only in a zone extending from approximately 30° N to 30° S of the equator. Tropical corals do not grow at depths of over 50 meters (160 ft). The optimum temperature for most coral reefs is 26–27 °C (79–81 °F), and few reefs exist in waters below 18 °C (64 °F). However, reefs in the Persian Gulf have adapted to temperatures of 13 °C (55 °F) in winter and 38 °C (100 °F) in summer. There are 37 species of scleractinian corals identified in such harsh environment around Larak Island.

Deep-water coral can exist at greater depths and colder temperatures at much higher latitudes, as far north as Norway. Although deep water corals can form reefs, very little is known about them.

Coral reefs are rare along the west coasts of the Americas and Africa, due primarily to upwelling and strong cold coastal currents that reduce water temperatures in these areas (respectively the Peru, Benguela and Canary streams). Corals are seldom found along the coastline of South Asia—from the eastern tip of India (Chennai) to the Bangladesh and Myanmar borders—as well as along the coasts of northeastern South America and Bangladesh, due to the freshwater release from the Amazon and Ganges Rivers respectively.

- The Great Barrier Reef—largest, comprising over 2,900 individual reefs and 900 islands stretching for over 2,600 kilometers (1,600 mi) off Queensland, Australia
- The Mesoamerican Barrier Reef System—second largest, stretching 1,000 kilometers (620 mi) from Isla Contoy at the tip of the Yucatán Peninsula down to the Bay Islands of Honduras
- The New Caledonia Barrier Reef—second longest double barrier reef, covering 1,500 kilometers (930 mi)
- The Andros, Bahamas Barrier Reef—third largest, following the east coast of Andros Island, Bahamas, between Andros and Nassau
- The Red Sea—includes 6000-year-old fringing reefs located around a 2,000 km (1,240 mi) coastline
- The Florida Reef Tract—largest continental US reef, extends from Soldier Key, located in Biscayne Bay, to the Dry Tortugas in the Gulf of Mexico
- Pulley Ridge—deepest photosynthetic coral reef, Florida
- Numerous reefs scattered over the Maldives
- The Philippines coral reef area, the second largest in Southeast Asia, is estimated at 26,000 square kilometers and holds an extraordinary diversity of species. Scientists have identified 915 reef fish species and more than 400 scleractinian coral species, 12 of which are endemic.
- The Raja Ampat Islands in Indonesia›s West Papua province offer the highest known marine diversity.

- Bermuda is known for its northernmost coral reef system, located at 32.4° N and 64.8° W. The presence of coral reefs at this high latitude is due to the proximity of the Gulf Stream. Bermuda has a fairly consistent diversity of coral species, representing a subset of those found in the greater Caribbean.
- The world›s northernmost individual coral reef so far discovered is located within a bay of Japan›s Tsushima Island in the Korea Strait.
- The world›s southernmost coral reef is at Lord Howe Island, in the Pacific Ocean off the east coast of Australia.

Biology

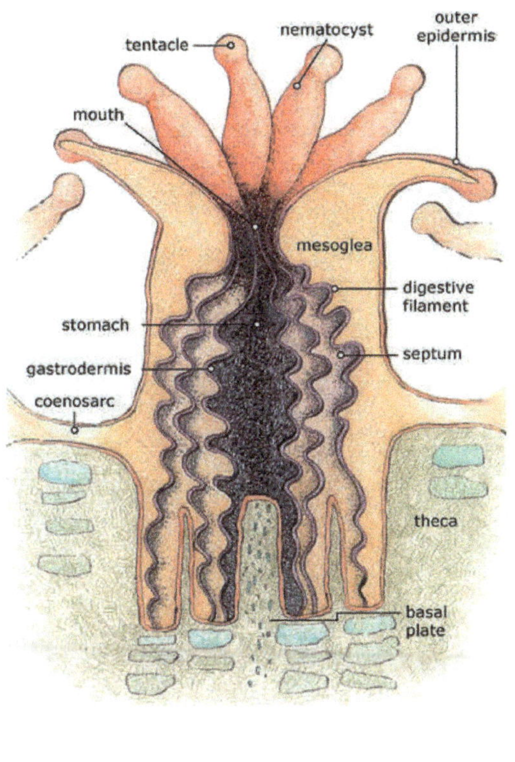

Anatomy of a coral polyp

Alive corals are colonies of small animals embedded in calcium carbonate shells. It is a mistake to think of coral as plants or rocks. Coral heads consist of accumulations of individual animals called polyps, arranged in diverse shapes. Polyps are usually tiny, but they can range in size from a pinhead to 12 inches (30 cm) across.

Reef-building or hermatypic corals live only in the photic zone (above 50 m), the depth to which sufficient sunlight penetrates the water, allowing photosynthesis to occur. Coral polyps do not photosynthesize, but have a symbiotic relationship with microscopic algae of the genus *Symbiodinium*, commonly referred to as zooxanthellae. These organisms live within the tissues of polyps and provide organic nutrients that nourish the polyp.

Because of this relationship, coral reefs grow much faster in clear water, which admits more sunlight. Without their symbionts, coral growth would be too slow to form significant reef structures. Corals get up to 90% of their nutrients from their symbionts.

Reefs grow as polyps and other organisms deposit calcium carbonate, the basis of coral, as a skeletal structure beneath and around themselves, pushing the coral head's top upwards and outwards.Waves, grazing fish (such as parrotfish), sea urchins, sponges, and other forces and organisms act as bioeroders, breaking down coral skeletons into fragments that settle into spaces in the reef structure or form sandy bottoms in associated reef lagoons. Many other organisms living in the reef community contribute skeletal calcium carbonate in the same manner. Coralline algae are important contributors to reef structure in those parts of the reef subjected to the greatest forces by waves (such as the reef front facing the open ocean). These algae strengthen the reef structure by depositing limestone in sheets over the reef surface.

Typical shapes for coral species are wrinkled brains, cabbages, table tops, antlers, wire strands and pillars. These shapes can depend on the life history of the coral, like light exposure and wave action, and events such as breakages.

Table coral

Close up of polyps are arrayed on a coral, waving their tentacles. There can be thousands of polyps on a single coral branch.

Corals reproduce both sexually and asexually. An individual polyp uses both reproductive modes within its lifetime. Corals reproduce sexually by either internal or external fertilization. The reproductive cells are found on the mesenteries, membranes that radiate inward from the layer of tissue that lines the stomach cavity. Some mature adult corals are hermaphroditic; others are exclusively male or female. A few species change sex as they grow.

Internally fertilized eggs develop in the polyp for a period ranging from days to weeks. Subsequent development produces a tiny larva, known as a planula. Externally fertilized eggs develop during synchronized spawning. Polyps release eggs and sperm into the water en masse, simultaneously. Eggs disperse over a large area. The timing of spawning depends on time of year, water temperature, and tidal and lunar cycles. Spawning is most successful when there is little variation between high and low tide. The less water movement, the better the chance for fertilization. Ideal timing occurs in the spring. Release of eggs or planula usually occurs at night, and is sometimes in phase with the lunar cycle (three to six days after a full moon). The period from release to settlement lasts only a few days, but some planulae can survive afloat for several weeks. They are vulnerable to predation and environmental conditions. The lucky few planulae which successfully attach to substrate next confront competition for food and space.

There are eight clades of Symbiodinium phylotypes. Most research has been completed on the Symbiodinium clades A–D. Each one of the eight contributes their own benefits as well as less compatible attributes to the survival of their coral hosts. Each photosynthetic organism has a specific level of sensitivity to photodamage of compounds needed for survival, such as proteins. Rates of regeneration and replication determine the organism›s ability to survive. Phylotype A is found more in the shallow regions of marine waters. It is able to produce mycosporine-like amino acids that are UV resistant, using a derivative of glycerin to absorb the UV radiation and allowing them to become more receptive to warmer water temperatures. In the event of UV or thermal damage, if and when repair occurs, it will increase the likelihood of survival of the host and symbiont. This leads to the idea that, evolutionarily, clade A is more UV resistant and thermally resistant than the other clades.

Clades B and C are found more frequently in the deeper water regions, which may explain the higher susceptibility to increased temperatures. Terrestrial plants that receive less sunlight because they are found in the undergrowth can be analogized to clades B, C, and D. Since clades B through D are found at deeper depths, they require an elevated light absorption rate to be able to synthesize as much energy. With elevated absorption rates at UV wavelengths, the deeper occurring phylotypes are more prone to coral bleaching versus the more shallow clades. Clade D has been observed to be high temperature-tolerant, and as a result it has a higher rate of survival than clades B and C.

Brain coral

Staghorn coral

Spiral wire coral

Pillar coral

Mushroom coral

Maze coral

Black coral

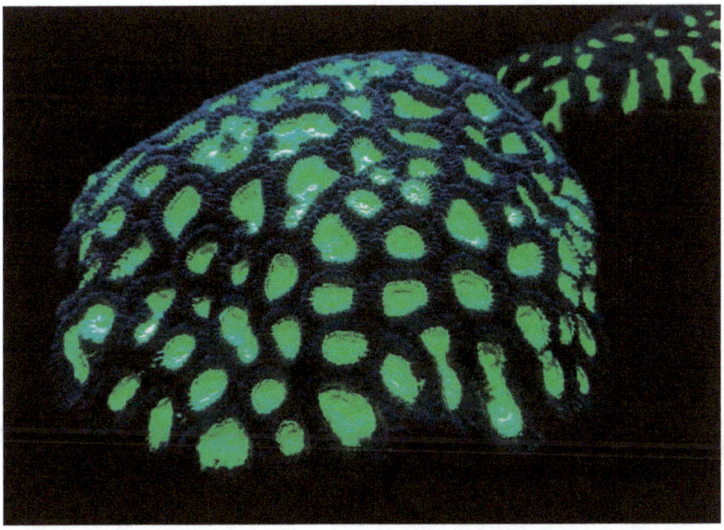

Fluorescent coral

Darwin's Paradox

In *The Structure and Distribution of Coral Reefs*, published in 1842, Darwin described how coral reefs were found in some areas of the tropical seas but not others, with no obvious cause. The largest and strongest corals grew in parts of the reef exposed to the most violent surf and corals were weakened or absent where loose sediment accumulated.

Tropical waters contain few nutrients yet a coral reef can flourish like an "oasis in the desert". This has given rise to the ecosystem conundrum, sometimes called "Darwin's paradox": "How can such high production flourish in such nutrient poor conditions?"

Coral reefs cover less than 0.1% of the surface of the world's ocean, about half the land area of France, yet they support over one-quarter of all marine species. This diversity results in complex food webs, with large predator fish eating smaller forage fish that eat yet smaller zooplankton and so on. However, all food webs eventually depend on plants, which are the primary producers. Coral reefs' primary productivity is very high, typically producing 5–10 grams of carbon per square meter per day ($gC{\cdot}m^{-2}{\cdot}day^{-1}$) biomass.

One reason for the unusual clarity of tropical waters is they are deficient in nutrients and drifting plankton. Further, the sun shines year-round in the tropics, warming the surface layer, making it less dense than subsurface layers. The warmer water is separated from deeper, cooler water by a stable thermocline, where the temperature makes a rapid change. This keeps the warm surface waters floating above the cooler deeper waters. In most parts of the ocean, there is little exchange between these layers. Organisms that die in aquatic environments generally sink to the bottom, where they decompose, which releases nutrients in the form of nitrogen (N), phosphorus (P) and potassium (K). These nutrients are necessary for plant growth, but in the tropics, they do not directly return to the surface.

Plants form the base of the food chain, and need sunlight and nutrients to grow. In the ocean, these plants are mainly microscopic phytoplankton which drift in the water column. They need sunlight for photosynthesis, which powers carbon fixation, so they are found only relatively near the surface. But they also need nutrients. Phytoplankton rapidly use nutrients in the surface waters, and in the tropics, these nutrients are not usually replaced because of the thermocline.

Coral polyps

Explanations

Around coral reefs, lagoons fill in with material eroded from the reef and the island. They become havens for marine life, providing protection from waves and storms.

Most importantly, reefs recycle nutrients, which happens much less in the open ocean. In coral reefs and lagoons, producers include phytoplankton, as well as seaweed and coralline algae, especially small types called turf algae, which pass nutrients to corals. The phytoplankton are eaten by fish and crustaceans, who also pass nutrients along the food web. Recycling ensures fewer nutrients are needed overall to support the community.

Coral reefs support many symbiotic relationships. In particular, zooxanthellae provide energy to coral in the form of glucose, glycerol, and amino acids. Zooxanthellae can provide up to 90% of a coral's energy requirements. In return, as an example of mutualism, the corals shelter the zooxanthellae, averaging one million for every cubic centimeter of coral, and provide a constant supply of the carbon dioxide they need for photosynthesis.

The color of corals depends on the combination of brown shades provided by their zooxanthellae and pigmented proteins (reds, blues, greens, etc.) produced by the corals themselves.

Corals also absorb nutrients, including inorganic nitrogen and phosphorus, directly from water. Many corals extend their tentacles at night to catch zooplankton that brush them when the water is agitated. Zooplankton provide the polyp with nitrogen, and the polyp shares some of the nitrogen with the zooxanthellae, which also require this element. The varying pigments in different species of zooxanthellae give them an overall brown or golden-brown appearance, and give brown corals their colors. Other pigments such as reds, blues, greens, etc. come from colored proteins made by the coral animals. Coral which loses a large fraction of its zooxanthellae becomes white (or sometimes pastel shades in corals that are richly pigmented with their own colorful proteins) and is said to be bleached, a condition which, unless corrected, can kill the coral.

Sponges are another key: they live in crevices in the coral reefs. They are efficient filter feeders, and in the Red Sea they consume about 60% of the phytoplankton that drifts by. The sponges eventually excrete nutrients in a form the corals can use.

Most coral polyps are nocturnal feeders. Here, in the dark, polyps have extended their tentacles to feed on zooplankton.

The roughness of coral surfaces is the key to coral survival in agitated waters. Normally, a boundary layer of still water surrounds a submerged object, which acts as a barrier. Waves breaking on the extremely rough edges of corals disrupt the boundary layer, allowing the corals access to passing nutrients. Turbulent water thereby promotes reef growth and branching. Without the nutritional gains brought by rough coral surfaces, even the most effective recycling would leave corals wanting in nutrients.

Studies have shown that deep nutrient-rich water entering coral reefs through isolated events may have significant effects on temperature and nutrient systems. This water movement disrupts the relatively stable thermocline that usually exists between warm shallow water to deeper colder water. Leichter et al. (2006) found that temperature regimes on coral reefs in the Bahamas and Florida were highly variable with temporal scales of minutes to seasons and spatial scales across depths.

Water can be moved through coral reefs in various ways, including current rings, surface waves, internal waves and tidal changes. Movement is generally created by tides and wind. As tides interact with varying bathymetry and wind mixes with surface water, internal waves are created. An internal wave is a gravity wave that moves along density stratification within the ocean. When a water parcel encounters a different density it will oscillate and create internal waves. While internal waves generally have a lower frequency than surface waves, they often form as a single wave that breaks into multiple waves as it hits a slope and moves upward. This vertical break up of internal waves causes significant diapycnal mixing and turbulence. Internal waves can act as nutrient pumps, bringing plankton and cool nutrient-rich water up to the surface.

The irregular structure characteristic of coral reef bathymetry may enhance mixing and produce pockets of cooler water and variable nutrient content. Arrival of cool, nutrient-rich water from depths due to internal waves and tidal bores has been linked to growth rates of suspension feeders and benthic algae as well as plankton and larval organisms. Leichter et al. proposed that Codium isthmocladum react to deep water nutrient sources due to their tissues having different concentrations of nutrients dependent upon depth. Wolanski and Hamner noted aggregations of eggs, larval organisms and plankton on reefs in response to deep water intrusions. Similarly, as internal waves and bores move vertically, surface-dwelling larval organisms are carried toward the shore. This has significant biological importance to cascading effects of food chains in coral reef ecosystems and may provide yet another key to unlocking "Darwin's Paradox".

Cyanobacteria provide soluble nitrates for the reef via nitrogen fixation.

Coral reefs also often depend on surrounding habitats, such as seagrass meadows and mangrove forests, for nutrients. Seagrass and mangroves supply dead plants and animals which are rich in nitrogen and also serve to feed fish and animals from the reef by supplying wood and vegetation. Reefs, in turn, protect mangroves and seagrass from waves and produce sediment in which the mangroves and seagrass can root.

Biodiversity

Tube sponges attracting cardinal fishes, glassfishes and wrasses

Over 4,000 species of fish inhabit coral reefs.

Organisms can cover every square inch of a coral reef.

Coral reefs form some of the world's most productive ecosystems, providing complex and varied marine habitats that support a wide range of other organisms.Fringing reefs just below low tide level have a mutually beneficial relationship with mangrove forests at high tide level and sea grass meadows in between: the reefs protect the mangroves and seagrass from strong currents and waves that would damage them or erode the sediments in which they are rooted, while the mangroves and sea grass protect the coral from large influxes of silt, fresh water and pollutants. This level of variety in the environment benefits many coral reef animals, which, for example, may feed in the sea grass and use the reefs for protection or breeding.

Reefs are home to a large variety of animals, including fish, seabirds, sponges, cnidarians (which includes some types of corals and jellyfish), worms, crustaceans (including shrimp, cleaner shrimp, spiny lobsters and crabs), mollusks (including cephalopods), echinoderms (including starfish, sea urchins and sea cucumbers), sea squirts, sea turtles and sea snakes. Aside from humans, mammals are rare on coral reefs, with visiting cetaceans such as dolphins being the main exception. A few of these varied species feed directly on corals, while others graze on algae on the reef. Reef biomass is positively related to species diversity.

The same hideouts in a reef may be regularly inhabited by different species at different times of day. Nighttime predators such as cardinalfish and squirrelfish hide during the day, while damselfish, surgeonfish, triggerfish, wrasses and parrotfish hide from eels and sharks.[49]

Algae

Reefs are chronically at risk of algal encroachment. Overfishing and excess nutrient supply from onshore can enable algae to outcompete and kill the coral. Increased nutrient levels can be a result of sewage or chemical fertilizer runoff from nearby coastal developments. Runoff can carry nitrogen and phosphorus which promote excess algae growth. Algae can sometimes out-compete the coral for space. The algae can then smother the coral by decreasing the oxygen supply available to the reef. Decreased oxygen levels can slow down coral's calcification rates weakening the coral and leaving it more susceptible to disease and degradation. In surveys done around largely uninhabited US Pacific islands, algae inhabit a large percentage of surveyed coral locations. The algal population consists of turf algae, coralline algae, and macro algae.

Sponges

Sponges are essential for the functioning of the coral reef's ecosystem. Algae and corals in coral reefs produce organic material. This is filtered through sponges which convert this organic material into small particles which in turn are absorbed by algae and corals.

Fish

Over 4,000 species of fish inhabit coral reefs. The reasons for this diversity remain unclear. Hypotheses include the "lottery", in which the first (lucky winner) recruit to a territory is typically able to defend it against latecomers, "competition", in which adults compete for territory, and less-competitive species must be able to survive in poorer habitat, and "predation", in which population size is a function of postsettlement piscivore mortality. Healthy reefs can produce up to 35 tons of fish per square kilometer each year, but damaged reefs produce much less.

Invertebrates

Sea urchins, Dotidae and sea slugs eat seaweed. Some species of sea urchins, such as *Diadema antillarum*, can play a pivotal part in preventing algae from overrunning reefs. Nudibranchia and sea anemones eat sponges.

A number of invertebrates, collectively called cryptofauna, inhabit the coral skeletal substrate itself, either boring into the skeletons (through the process of bioerosion) or living in pre-existing voids and crevices. Those animals boring into the rock include sponges, bivalve mollusks, and sipunculans. Those settling on the reef include many other species, particularly crustaceans and polychaete worms.

Seabirds

Coral reef systems provide important habitats for seabird species, some endangered. For example, Midway Atoll in Hawaii supports nearly three million seabirds, including two-thirds (1.5 million) of the global population of Laysan albatross, and one-third of the global population of black-footed albatross. Each seabird species has specific sites on the atoll where they nest. Altogether, 17 species of seabirds live on Midway. The short-tailed albatross is the rarest, with fewer than 2,200 surviving after excessive feather hunting in the late 19th century.

Other

Sea snakes feed exclusively on fish and their eggs. Marine birds, such as herons, gannets, pelicans and boobies, feed on reef fish. Some land-based reptiles intermittently associate with reefs, such as monitor lizards, the marine crocodile and semiaquatic snakes, such as *Laticauda colubrina*. Sea turtles, particularly hawksbill sea turtles, feed on sponges.

Schooling reef fish

Caribbean reef squid

Banded coral shrimp

reef shark

Green turtle

Giant clam

Soft coral, cup coral, sponges and ascidians

Banded sea krait

The shell of Latiaxis wormaldi, a coral snail

Importance

Coral reefs deliver ecosystem services to tourism, fisheries and coastline protection. The global economic value of coral reefs has been estimated to be between US $29.8 billion and $375 billion per year. Coral reefs protect shorelines by absorbing wave energy, and many small islands would not exist without their reefs to protect them. According to the environmental group World Wide Fund for Nature, the economic cost over a 25-year period of destroying one kilometer of coral reef is somewhere between $137,000 and $1,200,000. About six million tons of fish are taken each year from coral reefs. Well-managed coral reefs have an annual yield of 15 tons of seafood on average per square kilometer. Southeast Asia's coral reef fisheries alone yield about $2.4 billion annually from seafood.

To improve the management of coastal coral reefs, another environmental group, the World Resources Institute (WRI) developed and published tools for calculating the value of coral reef-related tourism, shoreline protection and fisheries, partnering with five Caribbean countries. As of April 2011, published working papers covered St. Lucia, Tobago, Belize, and the Dominican Republic, with a paper for Jamaica in preparation. The WRI was also "making sure that the study results support improved coastal policies and management planning". The Belize study estimated the value of reef and mangrove services at $395–559 million annually.

Bermuda's coral reefs provide economic benefits to the Island worth on average $722 million per year, based on six key ecosystem services, according to Sarkis *et al* (2010).

Threats

Island with fringing reef off Yap, Micronesia

Coral reefs are dying around the world. In particular, coral mining, agricultural and urban runoff, pollution (organic and inorganic), overfishing, blast fishing, disease, and the digging of canals and access into islands and bays are localized threats to coral ecosystems. Broader threats are sea temperature rise, sea level rise and pH changes from ocean acidification, all associated with greenhouse gas emissions. A 2014 study lists factors such as population explosion along the coast lines, overfishing, the pollution of coastal areas, global warming and invasive species among the main reasons that have put reefs in danger of extinction.

A study released in April 2013 has shown that air pollution can also stunt the growth of coral reefs; researchers from Australia, Panama and the UK used coral records (between 1880 and 2000) from the western Caribbean to show the threat of factors such as coal-burning coal and volcanic eruptions. Pollutants, such as Tributyltin, a biocide released into water from in anti-fouling paint can be toxic to corals.

In 2011, researchers suggested that "extant marine invertebrates face the same synergistic effects of multiple stressors" that occurred during the end-Permian extinction, and that genera "with poorly buffered respiratory physiology and calcareous shells", such as corals, were particularly vulnerable.

Rock coral on seamounts across the ocean are under fire from bottom trawling. Reportedly up to 50% of the catch is rock coral, and the practice transforms coral structures to rubble. With it taking years to regrow, these coral communities are disappearing faster than they can sustain themselves.

Another cause for the death of coral reefs is bioerosion. Various fishes graze corals, dead or alive and change the morphology of coral reefs making them more susceptible to other physical and chemical threats. It has been generally observed that only the algae growing on dead corals is eaten and the live ones are not. However, this act still destroys the top layer of coral substrate and makes it harder for the reefs to sustain.

In El Niño-year 2010, preliminary reports show global coral bleaching reached its worst level since another El Niño year, 1998, when 16% of the world's reefs died as a result of increased water temperature. In Indonesia's Aceh province, surveys showed some 80% of bleached corals died. Scientists do not yet understand the long-term impacts of coral bleaching, but they do know that bleaching leaves corals vulnerable to disease, stunts their growth, and affects their reproduction, while severe bleaching kills them. In July, Malaysia closed several dive sites where virtually all the corals were damaged by bleaching. To find answers for these problems, researchers study the various factors that impact reefs. The list includes the ocean's role as a carbon dioxide sink, atmospheric changes, ultraviolet light, ocean acidification, viruses, impacts of dust storms carrying agents to far-flung reefs, pollutants, algal blooms and others. Reefs are threatened well beyond coastal areas. Coral reefs with one type of zooxanthellae are more prone to bleaching than are reefs with another, more hardy, species.

General estimates show approximately 10% of the world's coral reefs are dead. About 60% of the world's reefs are at risk due to destructive, human-related activities. The threat to the health of reefs is particularly high in Southeast Asia, where 95% of reefs are at risk from local threats. By the 2030s, 90% of reefs are expected to be at risk from both human activities and climate change; by 2050, all coral reefs will be in danger.

Current research is showing that ecotourism in the Great Barrier Reef is contributing to coral disease, and that chemicals in sunscreens may contribute to the impact of viruses on zooxanthellae.

Protection

A diversity of corals

Marine protected areas (MPAs) have become increasingly prominent for reef management. MPAs promote responsible fishery management and habitat protection. Much like national parks and wildlife refuges, and to varying degrees, MPAs restrict potentially damaging activities. MPAs encompass both social and biological objectives, including reef restoration, aesthetics, biodiversity, and economic benefits. However, there are very few MPAs that have actually made a substantial difference. Research in Indonesia, Philippines and Papua New Guinea shows that there is no significant difference between an MPA site and an unprotected site. Conflicts surrounding MPAs involve lack of participation, clashing views of the government and fisheries, effectiveness of the area, and funding. In some situations, as in the Phoenix Islands Protected Area, MPAs can also provide revenue, potentially equal to the income they would have generated without controls, as Kiribati did for its Phoenix Islands.

According to the *Caribbean Coral Reefs - Status Report 1970-2012* made by the IUCN. States that; stopping overfishing especially key fishes to coral reef like parrotfish, coastal zone management which reduce human pressure on reef, (for example restricting the coastal settlement, development and tourism in coastal reef) and controlling pollution specially sewage wastage, may not only reduce coral declining but also reverse it and may let to coral reef more adaptable to changes relates to climate and acidification. The report shows that healthier reef in the Caribbean are those with large population of parrotfish in countries which protect these key fishes and sea urchins, banning fish trap and Spearfishing creating "resilient reefs". To help combat ocean acidification, some laws are in place to reduce greenhouse gases such as carbon dioxide. The Clean Water

Act puts pressure on state government agencies to monitor and limit runoff of pollutants that can cause ocean acidification. Stormwater surge preventions are also in place, as well as coastal buffers between agricultural land and the coastline. This act also ensures that delicate watershed ecosystems are intact, such as wetlands. The Clean Water Act is funded by the federal government, and is monitored by various watershed groups. Many land use laws aim to reduce CO_2 emissions by limiting deforestation. Deforestation causes erosion, which releases a large amount of carbon stored in the soil, which then flows into the ocean, contributing to ocean acidification. Incentives are used to reduce miles traveled by vehicles, which reduces the carbon emissions into the atmosphere, thereby reducing the amount of dissolved CO_2in the ocean. State and federal governments also control coastal erosion, which releases stored carbon in the soil into the ocean, increasing ocean acidification.High-end satellite technology is increasingly being employed to monitor coral reef conditions.

Biosphere reserve, marine park, national monument and world heritage status can protect reefs. For example, Belize's barrier reef, Sian Ka'an, the Galapagos islands, Great Barrier Reef, Henderson Island, Palau and Papahānaumokuākea Marine National Monument are world heritage sites.

In Australia, the Great Barrier Reef is protected by the Great Barrier Reef Marine Park Authority, and is the subject of much legislation, including a biodiversity action plan. They have compiled a Coral Reef Resilience Action Plan. This detailed action plan consists of numerous adaptive management strategies, including reducing our carbon footprint, which would ultimately reduce the amount of ocean acidification in the oceans surrounding the Great Barrier Reef. An extensive public awareness plan is also in place to provide education on the "rainforests of the sea" and how people can reduce carbon emissions, thereby reducing ocean acidification.

Inhabitants of Ahus Island, Manus Province, Papua New Guinea, have followed a generations-old practice of restricting fishing in six areas of their reef lagoon. Their cultural traditions allow line fishing, but no net or spear fishing. The result is both the biomass and individual fish sizes are significantly larger than in places where fishing is unrestricted.

Restoration

Coral aquaculture, also known as coral farming or coral gardening, is showing promise as a potentially effective tool for restoring coral reefs, which have been declining around the world. The process bypasses the early growth stages of corals when they are most at risk of dying. Coral seeds are grown in nurseries, then replanted on the reef. Coral is farmed by coral farmers who live locally to the reefs and farm for reef conservation or for income.

Efforts to expand the size and number of coral reefs generally involve supplying substrate to allow more corals to find a home. Substrate materials include discarded vehicle tires, scuttled ships, subway cars, and formed concrete, such as reef balls. Reefs also grow unaided on marine structures such as oil rigs. In large restoration projects, propagated hermatypic coral on substrate can be secured with metal pins, superglue or milliput. Needle and thread can also attach A-hermatype coral to substrate.

Coral fragments growing on nontoxic concrete

A substrate for growing corals referred to as Biorock is produced by running low voltage electrical currents through seawater to crystallize dissolved minerals onto steel structures. The resultant white carbonate (aragonite) is the same mineral that makes up natural coral reefs. Corals rapidly colonize and grow at accelerated rates on these coated structures. The electrical currents also accelerate formation and growth of both chemical limestone rock and the skeletons of corals and other shell-bearing organisms. The vicinity of the anode and cathode provides a high-pH environment which inhibits the growth of competitive filamentous and fleshy algae. The increased growth rates fully depend on the accretion activity.

During accretion, the settled corals display an increased growth rate, size and density, but after the process is complete, growth rate and density return to levels comparable to natural growth, and are about the same size or slightly smaller.

One case study with coral reef restoration was conducted on the island of Oahu in Hawaii. The University of Hawaii has come up with a Coral Reef Assessment and Monitoring Program to help relocate and restore coral reefs in Hawaii. A boat channel on the island of Oahu to the Hawaii Institute of Marine Biology was overcrowded with coral reefs. Also, many areas of coral reef patches in the channel had been damaged from past dredging in the channel. Dredging covers the existing corals with sand, and their larvae cannot build and thrive on sand; they can only build on to existing reefs. Because of this, the University of Hawaii decided to relocate some of the coral reef to a different transplant site. They transplanted them with the help of the United States Army divers, to a relocation site relatively close to the channel. They observed very little, if any, damage occurred to any of the colonies while they were being transported, and no mortality of coral reefs has been observed on the new transplant site, but they will be continuing to monitor the new transplant site to see how potential environmental impacts (i.e. ocean acidification) will harm the overall reef mortality rate. While trying to attach the coral to the new transplant site, they found the coral placed on hard rock is growing considerably well, and coral was even growing on the wires that attached the transplant corals to the transplant site. This gives new hope to future research on coral reef transplant sites. As a result of this coral restoration project, no environmental effects were seen from the transplantation process, no recreational activities were decreased, and no scenic areas were affected by the project. This is a great example that coral transplantation and restoration can work and thrive under the right conditions, which means there may be hope for other damaged coral reefs.

Another possibility for coral restoration is gene therapy. Through infecting coral with genetically modified bacteria, it may be possible to grow corals that are more resistant to climate change and other threats.

Reefs in the Past

Ancient coral reefs

Throughout Earth history, from a few thousand years after hard skeletons were developed by marine organisms, there were almost always reefs. The times of maximum development were in the Middle Cambrian (513–501 Ma), Devonian (416–359 Ma) and Carboniferous (359–299 Ma), owing to order Rugosa extinct corals, and Late Cretaceous (100–66 Ma) and all Neogene (23 Ma–present), owing to order Scleractinia corals.

Not all reefs in the past were formed by corals: those in the Early Cambrian (542–513 Ma) resulted from calcareous algae and archaeocyathids (small animals with conical shape, probably related to sponges) and in the Late Cretaceous (100–66 Ma), when there also existed reefs formed by a group of bivalves called rudists; one of the valves formed the main conical structure and the other, much smaller valve acted as a cap.

Measurements of the oxygen isotopic composition of the aragonitic skeleton of coral reefs, such as Porites, can indicate changes in the sea surface temperature and sea surface salinity conditions of the ocean during the growth of the coral. This technique is often used by climate scientists to infer the paleoclimate of a region.

CORAL REEF FISH

Coral reef fish are fish which live amongst or in close relation to coral reefs. Coral reefs form complex ecosystems with tremendous biodiversity. Among the myriad inhabitants, the fish stand out as colourful and interesting to watch. Hundreds of species can exist in a small area of a healthy reef, many of them hidden or well camouflaged. Reef fish have developed many ingenious specialisations adapted to survival on the reefs.

Coral reefs occupy less than one percent of the surface area of the world oceans, but still they provide a home for 25 percent of all marine fish species. Reef habitats are a sharp contrast to the open water habitats that make up the other 99% of the world oceans.

However, loss and degradation of coral reef habitat, increasing pollution, and overfishing including the use of destructive fishing practices, are threatening the survival of the coral reefs and the associated reef fish.

The fish that inhabit coral reefs are numerous and diverse.

Overview

Coral reefs are the result of millions of years of coevolution among algae, invertebrates and fish. They have become crowded and complex environments, and the fish have evolved many ingenious ways of surviving.Most fishes found on coral reefs are ray-finned fishes, known for the characteristic sharp, bony rays and spines in their fins. These spines provide formidable defences, and when erected they can usually be locked in place or are venomous. Many reef fish have also evolved cryptic coloration to confuse predators.

In the foreground is an orange-lined triggerfish displaying spines. Triggerfish have mouths that crush shells. Orange-lined triggerfish are particularly aggressive. The black and white fish are three-stripe damselfish and the unstriped fish are blue-green chromis damselfish. If the triggerfish attacks, the damselfish will hide in the nearby pillar coral. If the triggerfish wants to hide, it will squeeze into a coral crevice and lock itself in place with its spines.

Reef fish have also evolved complex adaptive behaviours. Small reef fish get protection from predators by hiding in reef crevices or by shoaling and schooling. Many reef fish confine themselves to one small neighbourhood where every hiding place is known and can be immediately accessed. Others cruise the reefs for food in shoals, but return to a known area to hide when they are inactive. Resting small fish are still vulnerable to attack by crevice predators, so many fish, such as triggerfish, squeeze into a small hiding place and wedge themselves by erecting their spines.

As an example of the adaptations made by reef fish, the yellow tang is a herbivore which feeds on benthic turf algae. They also provide cleaner services to marine turtles, by removing algal growth from their shells. They do not tolerate other fish with the same colour or shape. When alarmed, the usually placid yellow tang can erect spines in its tail and slash at its opponent with rapid sideways movements.

Most coral reef fish have spines in their fins like this damselfish.

The usually placid yellow tang can erect spines in its tail and slash at its opponent with rapid sideways movements.

Diversity and Distribution

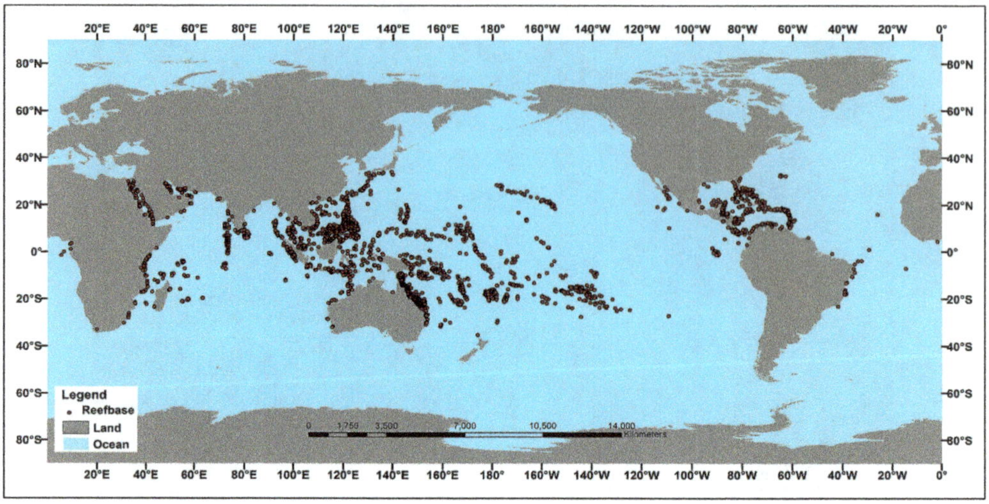

Distribution of coral reefs

Coral reefs contain the most diverse fish assemblages to be found anywhere on earth, with perhaps as many as 6,000-8,000 species that can be found dwelling within coral reef ecosystems of the world's oceans.

The mechanisms that first led to, and continue to maintain, such concentrations of fish species on coral reefs has been widely debated over the last 50 years. While many reasons have been proposed, there is no general scientific consensus on which of these is the most influential, but it seems likely that a number of factors contribute. These include the rich habitat complexity and diversity inherent in coral reef ecosystems, the wide variety and temporal availability of food resources available to coral reef fishes, a host of pre and post larval settlement processes, and as yet unresolved interactions between all these factors.

There are two major regions of coral reef development recognized; the Indo-Pacific (which includes the Pacific and Indian Oceans as well as the Red Sea), and the tropical western Atlantic (also known as the wider or greater Caribbean). Each of these two regions contains its own unique coral reef fish fauna with no natural overlap in species. Of the two regions, the richest by far in terms of reef fish diversity is the Indo-Pacific where there are an estimated 4,000-5,000 species of fishes associated with coral reef habitats. Another 500-700 species can be found in the greater Caribbean region.

Among goby species, small coral reef-dwelling fishes, is the world's shortest lived vertebrate, the seven-figure pygmy goby, which lives for less than 60 days.

The slowest-moving fishes are the sea horses, often found in reefs. The slowest of these, the dwarf seahorse, attains about five feet per hour.

Toadfish often inhabit reefs. Male toadfish "sing" at up to 100 decibels with their swim bladders to attract mates.

Reef Fish Adaptations

Contrast Between Coral Reef and Open Water Fishes

Many reef fish, like this queen angelfish, have a body flattened like a pancake, with pectoral and pelvic fins that act with the flattened body to maximize manoeuvrability.

In contrast, open water fish like this Atlantic bluefin tuna, are usually streamlined for straightline speed, with a deeply forked tail and a smooth body shaped like a spindle tapered at both ends. They are countershaded with silvery colours.

Body Shape

Most reef fishes have body shapes that are different from open water fishes. Open water fish are usually built for speed in the open sea, streamlined like torpedoes to minimise friction as they move through the water. Reef fish are operating in the relatively confined spaces and complex underwater landscapes of coral reefs. For this manoeuvrability is more important than straight line speed, so coral reef fish have developed bodies which optimize their ability to dart and change direction. They outwit predators by dodging into fissures in the reef or playing hide and seek around coral heads.

Many reef fish, such as butterflyfish and angelfishes, have evolved bodies which are deep and laterally compressed like a pancake. Their pelvic and pectoral fins are designed differently, so they act together with the flattened body to optimise manoeuvrability.

Colouration

The psychedelic mandarinfish is one of only two animal species known to have blue colouring because of cellular pigment.

The well camouflaged striated frogfish, a species of anglerfish, are ambush predators.

Clown triggerfish

Coral reef fishes exhibit a huge variety of dazzling and sometimes bizarre colours and patterns. This is in marked contrasts to open water fishes which are usually countershaded with silvery colours.

The patterns have different functions. Sometimes they camouflage the fish when the fish rests in places with the right background. Colouration can also be used to help species recognition during mating. Some unmistakable contrasting patterns are used to warn predators that the fish has venomous spines or poisonous flesh.

The foureye butterflyfish gets its name from a large dark spot on the rear portion of each side of the body. This spot is surrounded by a brilliant white ring, resembling an eyespot. A black vertical bar on the head runs through the true eye, making it hard to see. This can result in a predator thinking the fish is bigger than it is, and confusing the back end with the front end. The butterflyfish›s first instinct when threatened is to flee, putting the false eyespot closer to the predator than the head. Most predators aim for the eyes, and this false eyespot tricks the predator into believing that the fish will flee tail first. When escape is not possible, the butterflyfish will sometimes turn to face its aggressor, head lowered and spines fully erect, like a bull about to charge. This may serve to intimidate the other animal or may remind the predator that the butterflyfish is too spiny to make a comfortable meal.

The psychedelic mandarinfish (right) is not easily seen due to its bottom-feeding habit and its small size, reaching only about 6 cm. It feeds primarily on small crustaceans and other invertebrates, and is popular in the aquarium trade.

Just as some prey species evolved cryptic colouration and patterns to help avoid predators, some ambush predators evolved camouflage that lets them ambush their prey. The tassled scorpionfish is an ambush predator that looks like part of a sea floor encrusted with coral and algae. It lies in wait on the sea floor for crustaceans and small fish, such as gobies, to pass by. Another ambush predator is the striated frogfish (right). They lie on the bottom and wave a conspicuous worm-like lure strategically attached above their mouth. Normally about 10 cm (4 in) long, they can also inflate themselves like puffers.

Gobies avoid predators by tucking themselves into coral crevices or partly burying themselves in sand. They continually scan for predators with eyes that swivel independently. The camouflage of the tassled scorpionfish can prevent gobies from seeing them until it›s too late.

The clown triggerfish has strong jaws for crushing and eating sea urchins, crustaceans and hard-shell molluscs. Its ventral (lower) surface has large, white spots on a dark background, and its dorsal (upper) surface has black spots on yellow.This is a form of countershading: from below, the white spots look like the lighted surface of the water above; and from above, the fish blends more with the coral reef below. The brightly painted yellow mouth may deter potential predators.

The foureye butterflyfish has false eyes on its back end, confusing predators about which is the front end of the fish.

The frogfish is an ambush predator disguised to look like an algae-covered stone

Another ambush predator is the tassled scorpionfish camouflaged to look like part of a coral encrusted sea floor.

Gobies are very cautious, yet they can fail to see a tassled scorpionfish until it is too late.

Feeding Strategies

Many reef fish species have evolved different feeding strategies accompanied by specialized mouths, jaws and teeth particularly suited to deal with their primary food sources found in coral reef ecosystems. Some species even shift their dietary habits and distributions as they mature. This is not surprising, given the huge variety in the types of prey on offer around coral reefs.

For example, the primary food source of butterflyfishes are the coral polyps themselves or the appendages of polychaetes and other small invertebrate animals. Their mouths protrude like forceps, and are equipped with fine teeth that allow them to nip off such exposed body parts of their prey. Parrotfishes eat algae growing on reef surfaces, utilizing mouths like beaks well adapted to scrape off their food. Other fish, like snapper, are generalized feeders with more standard jaw and mouth structures that allow them to forage on a wide range of animal prey types, including small fishes and invertebrates.

Generalized Carnivores

Red snapper, are generalized reef feeders with standard jaw and mouth structures that allow them to eat almost anything, though they prefer small fish and crustaceans.

Carnivores are the most diverse of feeding types among coral reef fishes. There are many more carnivore species on the reefs than herbivores. Competition among carnivores is intense, resulting in a treacherous environment for their prey. Hungry predators lurk in ambush or patrol every part of the reef, night and day.

Some fishes associated with reefs are generalized carnivores that feed on a variety of animal prey. These typically have large mouths that can be rapidly expanded, thereby drawing in nearby water and any unfortunate animals contained within the inhaled water mass. The water is then expelled through the gills with the mouth closed, thereby trapping the helpless prey For example, the bluestripe snapper has a varied diet, feeding on fishes, shrimps, crabs, stomatopods, cephalopods and planktonic crustaceans, as well as plant and algae material. Diet varies with age, location and the prevalent prey items locally.

Goatfish are tireless benthic feeders, using a pair of long chemosensory barbels (whiskers) protruding from their chins to rifle through the sediments in search of a meal. Like goats, they seek anything edible: worms, crustaceans, molluscs and other small invertebrates are staples. The yellowfin goatfish (*Mulloidichthys vanicolensis*) often schools with the blue-striped snapper. The yellowfins change their colouration to match that of the snapper. Presumably this is for predator protection, since goatfish are a more preferred prey than bluestripe snapper. By night the schools disperse and individual goatfish head their separate ways to loot the sands. Other nocturnal feeders shadow the active goatfish, waiting patiently for overlooked morsels.

Moray eels and coral groupers (*Plectropomus pessuliferus*) are known to cooperate with each other when hunting. Grouper are protogynous hermaphrodites, who school in *harems* of three to fifteen females. When no male is available, in each school the largest female shifts sex to male. If the final male disappears, changes to the largest female occur, with male behavior occurring within several hours and sperm production occurring within ten days. It is probably as a result of behavioural triggers.

Adult coral trout hunt a variety of reef fish, particularly damselfish, while their juveniles mostly eat crustaceans such as prawns.

Bluestripe snapper eat just about anything.

Yellowfin goatfish change their colouration so they can school with the blue-striped snapper.

Coral grouper sometimes cooperate with giant morays in hunting.

Specialised Carnivores

Large schools of forage fish, such as surgeonfish and cardinalfish, move around the reef feeding on tiny zooplankton. The forage fish are, in turn, eaten by larger fish, such as the bigeye trevally. Fish receive many benefits from schooling behaviour, including defence against predators through better predator detection, since each fish is on the lookout. Schooling fish have developed remarkable displays of precise choreography which confuse and evade predators. For this they have evolved special pressure sensors along their sides, called lateral lines, that let them feel each other's movements and stay synchronized.

Bigeye trevally also form schools. They are swift predators who patrol the reef in hunting packs. When they find a school of forage fish, such as cardinalfish, they surround them and herd them close to the reef. This panics the prey fish, and their schooling becomes chaotic, leaving them open to attack by the trevally.

Many small reef fishes gain advantages by schooling.

Cardinalfish swim in schools for protection against trevally.

Bigeye trevally hunt cardinalfish in packs and herd them against the reef. When the cardinalfish panic and break school formation, the trevally pick them off.

The titan triggerfish can move relatively large rocks when feeding and is often followed by smaller fishes that feed on leftovers. They also use a jet of water to uncover sand dollars buried in sand.

Barracuda are ferocious predators on other fishes, with razor-sharp conical teeth which make it easy for them to rip their prey to shreds. Barracuda patrol the outer reef in large schools, and are extremely fast swimmers with streamlined, torpedo-shaped bodies.

Porcupinefish are medium to large sized, and are usually found swimming among or near coral reefs. They inflate their body by swallowing water, reducing potential predators to those with much bigger mouths.

Porcupinefish inflate themselves by swallowing water or air, which restricts potential predators to those with bigger mouths.

Fish can not groom themselves. Some fish specialise as cleaner fish, and establish cleaning stations where other fish can come to have their parasites nibbled away. The "resident fish doctor and dentist on the reef is the bluestreak cleaner wrasse". The bluestreak is marked with a conspicuous bright blue stripe and behaves in a stereotypical way which attracts larger fish to its cleaning station. As the bluestreak snacks on the parasites it gently tickles its client. This seems to bring the larger fish back again for regular servicing. The reef lizardfish secretes a mucus coating which reduces drag when they swim and also protects it from some parasites. But other parasites find the mucus itself good to eat. So lizardfish visit the cleaner wrasse, which clean the parasites from the skin, gills and mouth.

The Emperor angelfish feeds on coral sponges.

The titan triggerfish followed by small orange-lined triggerfish and moorish idol that feed on leftovers.

Two small cleaner wrasses servicing a larger fish at a cleaning station

The reef lizardfish secretes a mucus coating which reduces drag when they swim. But some parasites find the mucus good to eat

Herbivores

Surgeonfish are among the most common of coral reef herbivores, often feeding in shoals. This may be a mechanism for overwhelming the highly aggressive defence responses of small territorial damselfishes that vigorously guard small patches of algae on coral reefs.

Herbivores feed on plants. The four largest groups of coral reef fishes that feed on plants are the parrotfishes, damselfishes, rabbitfishes, and surgeonfishes. All feed primarily on microscopic and macroscopic algae growing on or near coral reefs.

Algae can drape reefs in kaleidoscopes of colours and shapes. Algae are primary producers, which means they are plants synthesising food directly from solar energy and carbon dioxide and other simple nutrient molecules. Without algae, everything on the reef would die. One important algal group, the bottom dwelling (benthic) algae, grows over dead coral and other inert surfaces, and provides grazing fields for herbivores such as parrotfish.

Parrotfish are named for their parrot-like beaks and bright colours. They are large herbivores that graze on the algae that grows on hard dead corals. Equipped with two pairs of crushing jaws and their beaks, they pulverize chunks of algae-coated coral, digesting the algae and excreting the coral as fine sand.

Smaller parrotfish are relatively defenceless herbivores, poorly defended against predators like barracuda. They have evolved to find protection by schooling, sometimes with other species like shoaling rabbitfish. Spinefoot rabbitfish are named for their defensive venomous spines, and they are seldom attacked by predators. Spines are a last-ditch defence. It is better to avoid predator detection in the first place, and avoid being thrust into risky spine-to-fang battles. So rabbitfish have also evolved skilful colour changing abilities.

Damselfish are a group of species that feed on zooplankton and algae, and are an important reef forage fish for larger predators. They are small, typically five centimetres (two inches) long. Many species are aggressive towards other fishes which also graze on algae, such as surgeonfish. Surgeonfish sometimes use schooling as a countermeasure to defensive attacks by solitary damselfish.

Relatively defenceless parrotfish feed on algae.

Ferocious barracuda prey in schools on parrotfish.

Coral rabbitfish have venomous spines which they erect if threatened.

Schooling spinefoot rabbitfish are often joined by defenceless parrotfish.

Symbiosis

A hawkfish, safely perched on Acropora, surveys its surroundings

Symbiosis refers to two species that have a close relationship with each other. The relationship can be mutualistic, when both species benefit from the relationship, commensalistic, when one species benefits and the other is unaffected, and parasitistic, when one species benefits, and the other is harmed.

An example of commensalism occurs between the hawkfish and fire coral. Thanks to their large, skinless pectoral fins, hawkfish can perch on fire corals without harm. Fire corals are not true corals, but are hydrozoans possessing stinging cells called nematocysts which would normally prevent close contact. The protection fire corals offer hawkfish means the hawkfish has the high ground of the reef, and can safely survey its surroundings like a hawk. Hawkfish usually stay motionless, but dart out and grab crustaceans and other small invertebrates as they pass by. They are mostly solitary, although some species form pairs and share a head of coral.

A more bizarre example of commensalism occurs between the slim, eel-shaped pinhead pearlfish and a particular species of sea cucumber. The pearlfish enters the sea cucumber through its anus, and spends the day safely protected inside the seacucumbers alimentary tract. At night it emerges the same way and feeds on small crustaceans.

Sea anemones are common on reefs. The tentacles of sea anemones bristle with tiny harpoons (nematocysts) primed with toxins, and are an effective deterrent against most predators. However, saddle butterflyfish, which are up to 30 cm (12 in) long, have developed a resistance to these toxins. Saddle butterflyfish usually flutter gently rather than swim. But in the presence of their preferred food, sea anemones, this gentleness disappears, and the butterflyfish dash in and out, ripping off the anemone tentacles.

The tentacles of sea anemones bristle with tiny toxic harpoons.

Saddle butterflyfish are resistant to the sea anemone toxin.

Yellowtail clownfish with sea anemone

Common clownfish guarding their sea anemone home

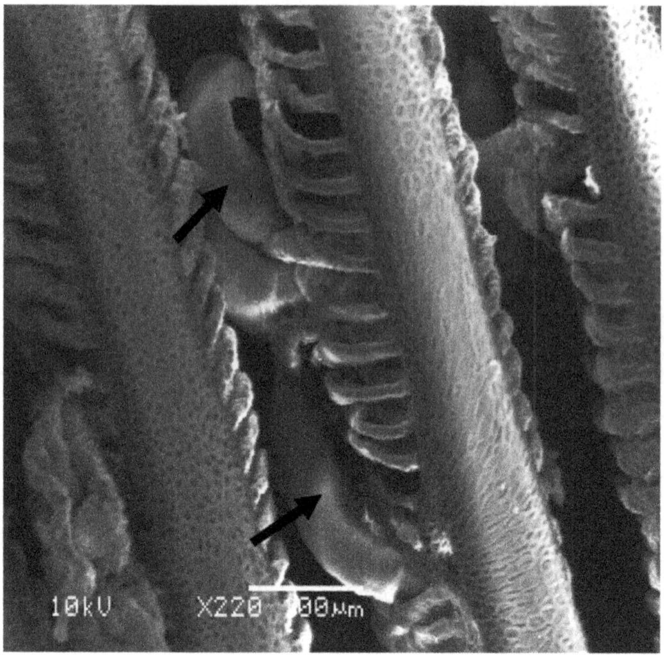

Monogenean parasites of the genus *Pseudorhabdosynochus* (arrows) on the gill filament of a grouper.

There is a mutualistic relationship between sea anemones and clownfish. This gives the sea anemones a second line of defence. They are guarded by fiercely territorial clownfish, who are also immune to the anemone toxins. To get their meal, butterflyfish must get past these protective clownfish who, although smaller, are not intimidated. An anemone without its clownfish will quickly be eaten by butterflyfish. In return, the anemones provide the clownfish protection from their predators, who are not immune to anemone stings. As a further benefit to the anemone, waste ammonia from the clownfish feed symbiotic algae found in the anemone›s tentacles.

As with all fish, coral reef fish harbour parasites. Since coral reef fish are characterized by high biodiversity, parasites of coral reef fish show tremendous variety. Parasites of coral reef fish include nematodes, Platyhelminthes (cestodes, digeneans, and monogeneans), leeches, parasitic crustaceans such as isopods and copepods, and various microorganisms such as myxosporidia and microsporidia. Some of these fish parasites have heteroxenous life cycles (i.e. they have several hosts) among which sharks (certain cestodes) or molluscs (digeneans). The high biodiversity of coral reefs increases the complexity of the interactions between parasites and their various and numerous hosts. Numerical estimates of parasite biodiversity have shown that certain coral fish species have up to 30 species of parasites. The mean number of parasites per fish species is about ten. This has a consequence in term of co-extinction. Results obtained for the coral reef fish of New Caledonia suggest that extinction of a coral reef fish species of average size would eventually result in the co-extinction of at least ten species of parasites.

Toxicity

Head on view of the beautiful but venomous lionfish

Many reef fish are toxic. Toxic fish are fish which contain strong toxins in their bodies. There is a distinction between poisonous fish and venomous fish. Both types of fish contain strong toxins, but the difference is in the way the toxin is delivered. Venomous fish deliver their toxins (called venom) by biting, stinging, or stabbing, causing an envenomation. Venomous fish don't necessarily cause poisoning if they are eaten, since the venom is often destroyed in the digestive system. By contrast, poisonous fish contain strong toxins which are not destroyed by the digestive system. This makes them poisonous to eat.

Venomous fish carry their venom in venom glands and use various delivery systems, such as spines or sharp fins, barbs or spikes, and fangs. Venomous fish tend to be either very visible, using flamboyant colours to warn enemies, or skilfully camouflaged and maybe buried in the sand. Apart from the defence or hunting value, venom might have value for bottom dwelling fish by killing the bacteria that try to invade their skin. Few of these venoms have been studied. They are a yet to be tapped resource for bioprospecting to find drugs with medical uses.

The most venomous known fish is the reef stonefish. It has a remarkable ability to camouflage itself amongst rocks. It is an ambush predator that sits on the bottom waiting for prey to come close. It does not swim away if disturbed, but erects 13 venomous spines along its back. For defence, it can shoot venom from each or all of these spines. Each spine is like a hypodermic needle, delivering the venom from two sacs attached to the spine. The stonefish has control over whether to shoot its venom, and does so when provoked or frightened. The venom results in severe pain, paralysis and tissue death, and can be fatal if not treated. Despite its formidable defence, the stonefish does have predators. Some bottom feeding rays and sharks with crushing teeth feed on them, as does the Stokes' seasnake

The most venomous known fish is the reef stonefish.

The spotted trunkfish secretes a ciguatera toxin from glands on its skin.

Like many other apex reef fish, the giant moray can cause ciguatera poisoning if eaten.

The stargazer buries itself in sand and can deliver electric shocks as well as venom.

Unlike the stonefish which can shoot venom, the lionfish can only release venom when something strikes its spines. Although not native to the US coast, lionfish have appeared around Florida and have spread up the coast to New York. They are attractive aquarium fish, sometimes used to stock ponds, and may have been washed into the sea during a hurricane. Lionfish can aggressively dart at scuba divers and attempt to puncture the facemask with their venomous spines.

The spotted trunkfish is a reef fish which secretes a colourless ciguatera toxin from glands on its skin when touched. The toxin is only dangerous when ingested, so there's no immediate harm to divers. However, predators as large as nurse sharks can die as a result of eating a trunkfish. Ciguatera toxins appear to accumulate in top predators of coral reefs. Many of the Caribbean groupers and the barracuda for example may contain enough of this toxin to cause severe symptoms in humans who eat them. What makes the situation particularly dangerous is that such species may be toxic only at certain sizes or locations, making it difficult to know whether or when they are or are not safe to eat. In some locations this leads to many cases of ciguatera poisoning among tropical islanders.

The stargazer buries itself in sand and can deliver electric shocks as well as venom. It is a delicacy in some cultures (the venom is destroyed when it is cooked), and can be found for sale in some fish markets with the electric organ removed. They have been called the meanest things in creation.

The giant moray is a reef fish at the top of the food chain. Like many other apex reef fish, it is likely to cause ciguatera poisoning if eaten. Outbreaks of ciguatera poisoning in the 11th to 15th centuries from large, carnivorous reef fish, caused by harmful algal blooms, could be a reason why Polynesians migrated to Easter Island, New Zealand, and possibly Hawaii.

Reef Sharks and Rays

The whitetip reef shark almost exclusively inhabits coral reefs.

Coral reefs in the Indo-Pacific are dominated by whitetip, blacktip and grey reef shark. Coral reefs in the western Atlantic Ocean are dominated by the Caribbean reef shark. These sharks are all species of requiem shark, and all have the robust, streamlined bodies that are typical of the requiem shark. They are fast-swimming, agile predators that feed primarily on free-swimming bony fishes and cephalopods. Other species of reef sharks include the Galapagos shark, the tawny nurse shark and hammerheads.

The whitetip reef shark is a small shark usually less than 1.6 m (5.2 ft) in length. It is associated almost exclusively with coral reefs where it can be encountered around coral heads and ledges with high vertical relief, or over sandy flats, in lagoons, or near drop-offs to deeper water. They prefer very clear water and rarely swim far from the bottom. Whitetip reef sharks spend most of the daytime time resting inside caves. Unlike other requiem sharks, which usually rely on ram ventilation and must constantly swim to breathe, this shark can pump water over its gills and lie still on the bottom. They have slender, lithe body, which allows them to wriggle into crevices and holes and extract prey inaccessible to other reef sharks. On the other hand, they are rather clumsy when attempting to take food suspended in open water.

Whitetip reef sharks spend much of the day lying still on the bottom.

The whitetip reef shark does not frequent very shallow water like the blacktip reef shark, nor the outer reef like the grey reef shark. They generally remain within a highly localized area. An individual shark may use the same cave for months to years. The daytime home range of a whitetip reef shark is limited to about 0.05 km^2 (0.019 sq mi), and at night this range increases to 1 km^2 (0.39 sq mi).

The whitetip reef shark is highly responsive to olfactory, acoustic, and electrical cues given off by potential prey. Its visual system is attuned more to movement and/or contrast than to object details. It is especially sensitive to natural and artificial low-frequency sounds in the 25–100 Hz range, which evoke struggling fish. They hunt primarily at night, when many fishes are asleep and easily taken. After dusk, a group of sharks may target the same prey item, covering every exit route from a particular coral head. Each shark hunts for itself and in competition with the others in its group. They feed mainly on bony fishes, including eels, squirrelfishes, snappers, damselfishes, parrotfishes, surgeonfishes, triggerfishes, and goatfishes, as well as octopus, spiny lobsters, and crabs. Important predators of the whitetip reef shark include tiger sharks and Galapagos sharks.

Adult blacktip reef sharks are often found patrolling reef ledges.

The blacktip reef shark is typically about 1.6 m (5.2 ft) long. It is usually found over reef ledges and sandy flats, though it can also enter brackish and freshwater environments. This species likes shallow water, while the whitetip and the grey reef shark are prefer deeper water. Younger sharks favour shallow sandy flats, and older sharks spend more time around reef ledges and near reef drop-offs. Blacktip reef sharks are strongly attached to their own area, where they may remain for up to several years. A tracking study off Palmyra Atoll in the central Pacific has found that the blacktip reef shark had a home range of about 0.55 km^2 (0.21 sq mi), among the smallest of any shark species. The size and location of the range does not change with time of day. The blacktip reef shark may be encountered alone or in small groups. Large social aggregations have also been observed. They are active predators of small bony fishes, cephalopods, and crustaceans, and also feed on sea snakes and seabirds. Blacktip reef sharks are preyed on by groupers, grey reef sharks, tiger sharks, and members of their own species. At Palmyra Atoll, adult blacktip reef sharks avoid patrolling tiger sharks by staying out of the central, deeper lagoon.

Grey reef sharks are usually less than 1.9 metres (6.2 ft) long. Despite their moderate size, grey reef sharks actively expel most other shark species from favored habitats. In areas where this species co-exists with the blacktip reef shark, the latter species occupy the shallow flats while the grey reef sharks stay in deeper water. Many grey reef sharks have a home range on a specific area of the reef, to which they continually return. However, they are social rather than territorial. During the day, these sharks often form groups of 5–20 individuals near coral reef drop-offs, splitting up in the evening as the sharks begin to hunt. They are found over continental and insular shelves, preferring the leeward (away from the direction of the current) sides of coral reefs with clear water and rugged topography. They are frequently found near the drop-offs at the outer edges of the reef, and less commonly within lagoons. On occasion, this shark may venture several kilometers out into the open ocean.

Coral reef drop-offs are favoured habitat for grey reef sharks.

Shark researcher Leonard Compagno comments on the relationship between the three species.

A Caribbean reef shark cruises a coral reef in the Bahamas.

\

The grey reef shark ...shows microhabitat separation from the blacktip reef sharks; around islands where both species occur, the blacktip occupies shallow flats, while the grey reef shark is usually found in deeper areas, but where the blacktip is absent, the grey reef shark is commonly found on the flats... The grey reef shark complements the whitetip shark as it is far more adapt at catching off-bottom fish than the whitetip, but the later is far more competent in extracting prey from crevices and holes in reefs.

The Caribbean reef shark is up to 3 metres (10 ft) long, one of the largest apex predators in the reef ecosystem. Like the whitetip reef shark, they have been documented resting motionless on the sea bottom or inside caves, unusual behaviour for requiem sharks. Caribbean reef shark play a major role in shaping Caribbean reef communities. They are more active at night, with no evidence of seasonal changes in activity or migration. Juveniles tend to remain in a localized area throughout the year, while adults range over a wider area. The Caribbean reef shark feeds on a wide variety of reef-dwelling bony fishes and cephalopods, as well as some elasmobranchs such as eagle rays and yellow stingrays . Young sharks feed on small fishes, shrimps, and crabs. In turn, young sharks are preyed on by larger sharks such as the tiger shark and the bull shark.

The great hammerhead uses its hammer both to locate electrical signatures of stingrays buried in the sand and to pin them down.

COASTAL FISH

Coastal fish, also called inshore fish or neritic fish, inhabit the sea between the shoreline and the edge of the continental shelf. Since the continental shelf is usually less than 200 metres deep, it follows that pelagic coastal fish are generally epipelagic fish, inhabiting the sunlit epipelagic zone. Coastal fish can be contrasted with oceanic fish or offshore fish, which inhabit the deep seas beyond the continental shelves.

Coastal fish are the most abundant in the world. They can be found in tidal pools, fjords and estuaries, near sandy shores and rocky coastlines, around coral reefs and on or above the continental shelf. Coastal fish include forage fish and the predator fish that feed on them. Forage fish thrive in inshore waters where high productivity results from upwelling and shoreline run off of nutrients. Some are partial residents that spawn in streams, estuaries and bays, but most complete their life cycles in the zone.

Schooling threadfin, a coastal species

Coastal Habitats

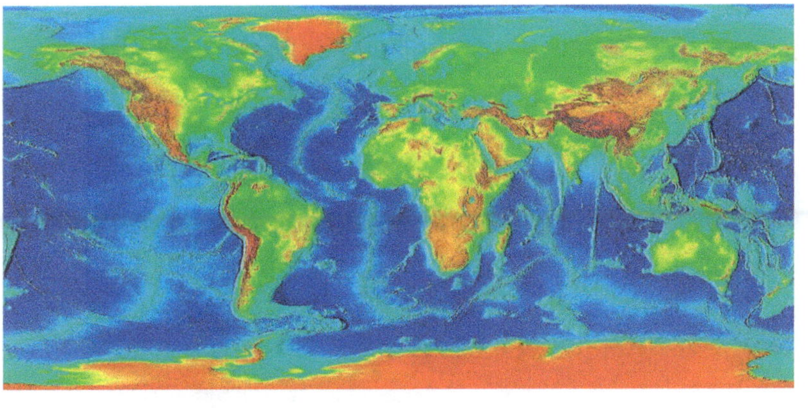

The global continental shelf, highlighted in cyan

Coastal fish are found in the waters above the continental shelves that extend from the continental shorelines, and around the coral reefs that surround volcanic islands. The total world shoreline extends for 356,000 km (221,000 mi) and the continental shelves occupy a total area of 24.286 million km^2 (9 376 million sq mi).This is about 4.8% of the world›s total area of 510.072 million km^2.

Nearshore Fish

Nearshore fish, sometimes called littoral fish, live close to the shore. They are associated with the intertidal zone, or with estuaries, lagoons, coral reefs, kelp forests, seagrass meadows, or rocky or sandy bottoms, usually in shallow waters less than about 10 m deep.

Intertidal Fish

Intertidal zones

The rise and fall of tides on a seashore defines an intertidal zone

Intertidal zones can be a volatile habitat for fish

Intertidal fish are fish that move in and out with the tide in the intertidal zone of the seashore, or are found in rock pools or under rocks.

The intertidal zone of rocky shores can contain indentations which trap pools of salty water, called rock pools. Living in these habitats are communities of hardy plant and animal species specially adapted for coping with the volatile environment around them. The plants and animals interact with each other and with the rock pool to form miniature ecosystems, easily accessible to students and a source of fascination for young children. Plants such as seaweeds, cnidarians such as sea anemones, arthropods like barnacles, and molluscs such as the common limpet and the common periwinkle can be permanent residents of rock pools. But most rock pool animals, such as crabs, shrimp and fish are just temporary residents, occupying a rock pool only until the next tide takes them to a new location.

Some rock pool fish which are temporary residents include the long-spined sea scorpion, the pipefish worm, the rock goby and the common lumpsucker. However some other rock pool fish are territorial in nature, and will stay with the same pool for extended periods. Examples are the common blenny and its near relative the butterfish.

- The common blenny, also known as the shanny, is found in northern temperate waters. They hide under rocks and in crannies in rock pools when the tide is out. They feed on green seaweed and invertebrates such as barnacles. They can crawl on dry land, using their paired fins. About 16 cm long, they have smooth skin, without scales, and are covered with soft slime. The slime prevents them drying if they are stranded on a shore between tides. So long as their skin stays moist, they can breathe out of water. They are sometimes called sea frogs because they bask in the sun on weeds outside the water, and like frogs, jump to safety when disturbed. They can change their colour to match their surroundings. The female lays eggs in crevices or under stones and the male guards them until they hatch. In the winter, when storms can be severe, they move out of their rock pools into the shallows. The common blenny is bold with strong teeth, and will bite humans if it feels threatened.
- The rock goby is a small fish, about 12 cm long, found in northern temperate waters. It is coloured black with white blotches, and hides under stones and amongst seaweed. It is a temporary resident of rock pools when the tide is out. The female rock goby lays eggs on the underside of rocks and shells and then leaves them. The male guards the eggs until they hatch. First-year rock gobies often visit rock pools in winter when the older fish have left.

Australian blenny

Rock goby

Long-spined sea scorpion

Smooth lumpsucker inflated in a defensive response.

- The long-spined sea scorpion, a small stout fish which grows about 29 cm long, is another temporary resident of rocky pools. They have large black eyes, a large mouth, and four long spines—two on each side on the gill cover—that stick out when the fish is removed from the water. They also have an organ like a finger on each side of their mouths which helps them catch prey. Because of their broad heads, they are also called bullheads. They have a variety of effective camouflaged colours ranging from shades of browns with cream blotches, to orange and red with white blotches. They can also change their body colour to match their surroundings. They are found around the coasts of Northern Europe in shallow rocky waters hiding amongst seaweed. They are also found in rock pools and sometimes in waters 30 m deep. Long-spined sea scorpions lay eggs amongst seaweed or attached to rock crevices. The young hatch after two or three weeks, and go through several development stages before maturing into adults.

- Lumpsuckers are found in temperate northern waters. They live on the seafloor, and are temporary residents of rocky pools in late winter and early spring when they spawn. The body of the lumpsucker is scaleless and covered with small lumps. They have a large sucking disc on their underside which they use to cling to surfaces. They are normally a blue to slate-grey colour, and are effectively camouflaged to look like stones. They are portly, nearly spherical, poor swimmers, reaching lengths up to 50 cm. After the female lumpsucker lays eggs, the male takes over, clamping itself to a rock where it guards the eggs. When they hatch, lumpsuckers look like tiny tadpoles. They remain in shallow water and rock pools, hiding amongst seaweed and rocks, until they grow up.

Estuarine Fish

This estuary of the Klamath River is a transition zone between a freshwater river environment and a saltwater marine environment. Due to land runoff, river mouths and estuary waters can be turbid and nutrient rich, sometimes to the point of eutrophication.

Estuaries are partly enclosed coastal bodies of water with one or more rivers or streams flowing into them, and with a free connection to the open sea. These brackish water habitats form a transition zone between river environments and ocean environments, and ecological successions can form along the way. Estuaries are subject to both marine influences, such as tides, waves, and the influx of saline water; and riverine influences, such as flows of fresh water and sediment. The inflow of both seawater and freshwater provide high levels of nutrients in both the water column and sediment, making estuaries productive natural habitats.

Fishes that spend time in estuaries (or river mouths) need to be euryhaline (tolerant to a range of salinities). Estuaries provide an unstable environment for fish, where the salinity changes and the waters are often muddy and turbulent. In warmer climates, estuaries have mangroves around their edges. At times there may be only a few different fish species present in an estuary, but seasonal migrants, including eels, salmonids, and some forage fish such as herrings and sprats increase the diversity in the estuary.

River estuaries form important staging points during the migration of anadromous and catadromus fish species, such as salmon and eels, giving them time to form social groups and to adjust to the changes in salinity. Salmon are anadromous, meaning they live in the sea but ascend rivers to spawn; eels are catadromous, living in rivers and streams, but returning to the sea to breed. Besides the species that migrate through estuaries, there are many other fish that use them as "nursery grounds" for spawning or as places young fish can feed and grow before moving elsewhere. For example, herring and plaice are two commercially important species that use the Thames Estuary for this purpose.

Mangrove swamps are associated brackish water habitats. Many, though not all, mangrove swamps fringe estuaries and lagoons where the salinity changes with each tide. Among the most specialised residents of mangrove forests are mudskippers, fish that forage for food on land, and archer fish, perch-like fish that spit at insects and other small animals living in the trees, knocking them into the water where they can be eaten. Like estuaries, mangrove swamps are important breeding grounds for many fish, with species such as snappers, halfbeaks, and tarpon spawning or maturing among them.

Coral Reef Fish

In tropical waters, coral reef fish live amongst or in close relation to coral reefs. Coral reefs form complex ecosystems with tremendous biodiversity. Coral reef fish can be particularly colourful and interesting to watch. Hundreds of species can exist in a small area of a healthy reef, many of them hidden or well camouflaged. Reef fish have developed many ingenious specialisations adapted to survival on the reefs. Coral reefs occupy less than one percent of the surface area of the world oceans, yet they provide a home for 25 percent of all marine fish species.

Coral reefs often depend on other habitats in the surrounding area for the supply of nutrients, such as seagrass meadows and mangrove forests. Seagrass and mangroves supply dead plants and animals which are rich in nitrogen and also serve to feed fish and animals from the reef by supplying wood and vegetation. Reefs in turn protect mangroves and seagrass from waves and produce sediment for the mangroves and seagrass to root in.

The sea goldie is an anthias. They are hermaphrodite, and swim in harems

Anthias are members of the family Serranidae (basses, basslets, groupers) and make up the subfamily Anthiinae. They are widespread in tropical waters. They have been called the quintessential reef fish, and make up a sizeable portion of the colourful fishes seen swarming in coral reef photography. Anthias are mostly small, peaceful, beautiful and popular as ornamental fish. They are mainly zooplankton feeders. Anthias shoal and school in large numbers, operating more intimate "harems" within the schools. These harems contain a dominant and colourful male, between 2 and 12 females — who operate a hierarchy among themselves — and one or two subdominant males, often less brightly coloured and non-territorial. Within the swarm of females, territorial males perform acrobatic U-swim displays and vigorously defend an area of the reef and its associated harem. Anthias are protogynous hermaphrodites. All anthias are born female; if a dominant male perishes, the largest female of the group will often change into a male to take its place. This may lead to squabbling between the next largest male and the transforming female, whose hormones are now surging with testosterone. This can turn quite vicious in the limited confines of captivity.

The foureye butterflyfish has a false eyespot on its sides, which can confuse prey and predators.

Butterflyfish are group of about 120 species belonging to the family Chaetodontidaeof Perchiformes. They include bannerfish and coralfish. They are widespread on coral reefs. Butterflyfish are mostly between 12 and 22 centimetres (4.7 and 8.7 in) in length. The largest species, the lined butterflyfish and saddle butterflyfish, grow to 30 centimetres (12 in). Many species are brightly coloured and strikingly patterned, though other species are dull in colour. Many have eyespots on their flanks and dark bands across their eyes, not unlike the patterns seen on butterfly wings.[:184] Their deep, laterally narrow bodies are easily noticed through the profusion of reef life. The conspicuous colouration of butterflyfish may be intended for interspecies communication. Butterflyfish have uninterrupted dorsal fins with tail fins that may be rounded or truncated, but are never forked. Generally diurnal and frequenting waters of less than 18 metres (59 ft) (though some species descend to 180 metres (590 ft)), butterflyfish stick to particular home ranges. The corallivores are especially territorial, forming mated pairs and staking claim to a specific coral head. Contrastingly, the zooplankton feeders form large conspecific groups. By night butterflyfish hide in reef crevices and exhibit markedly different colouration. Their colouration also makes butterflyfish popular aquarium fish. However, most species feed on coral polyps and sea anemones, which can result in problems for the hobby aquarists.

Clownfish, anemonefish and damselfish are among about 360 species classified in the family Pomacentridae. Most Pomacentrids are associated with coral reefs in the Indo-West Pacific, with a few species occurring in temperate waters. Some species are native to freshwater or brackish estuarine environments.[:205–208] Most live in shallow water, from 2 to 15 meters (6.6 to 49.2 ft), although some species are found below 100 m (330 ft). Most species are specialists, living in specific parts of the reef, such as sandy lagoons, steep reef slopes, or areas exposed to strong wave action. In general, the coral is used as shelter, and many species can only survive in its presence. The bottom-dwelling species are territorial, occupying and defending a portion of the reef, often centred around an area of shelter. By keeping away other species of fish, some pomacentrids encourage the growth of thick mats of algae within their territories, leading to the common name farmerfish. Different species display a wide range of colours, although some are relatively drab. Pomacentrids are omnivorous or herbivorous, feeding off algae, plankton, and small bottom-dwelling crustaceans. A small number eat coral.

The yellowfin goatfish changes its colour so it can school with blue-striped snappers

Goatfishes are a family Mullidae of about 55 species of perciform fishes, associated worldwide with tropical reefs.[186] They are typically about 20 cm long, though the dash-and-dot goatfish, grows to 55 cm. Goatfish are tireless benthic feeders, possess a pair of long chemosensory barbels ("whiskers") protruding from their chins resembling a goat's beard. They use these to rifle through the sediments in search of a meal. Like goats, they seek anything edible; worms, crustaceans, molluscs and other small invertebrates are staples. Many species of goatfish are conspicuously coloured and have the ability to change their colouration depending on their current activity. By day, many form large inactive (non-feeding) schools: these aggregates may contain both conspecifics and heterospecifics. For example, the yellowfin goatfish school with blue-striped snappers. When they do that, the yellowfins changes its colouration to match that of the snapper. By night the schools disperse and individual goatfish head their separate ways to loot the sands. The diurnal goldsaddle goatfish changes from a lemon-yellow to a pale cream when feeding. Other nocturnal feeders will shadow the active goatfish, waiting patiently for overlooked morsels. Goatfish stay within the shallows, going no deeper than about 110 metres. Most species do not tolerate brackish water, so they do not enter estuaries or the mouths of rivers.

Other Nearshore Fish

Other nearshore or shallow water fish live near the shore in depths of less than 10 metres. They occupy the areas over sandy or rocky bottoms, and can be associated with seagrass meadows and kelp forests. They can be divided into demersal fish and pelagic fish. Demersal fish live on or near the sea floor, while pelagic fish live in the water column away the sea floor.

Kelp forests can provide shelter and food for shallow water fish

Examples of such shallow water demersal fish, found in both tropical and temperate waters around the world, are triplefins, seahorses, wrasse and flounder. As demersal fish, all these fish spend most of their time on or near the sea floor.

- Flatfish are superbly adapted groundfish, found on muddy and sandy sea floors. In many species both eyes lie on one side of the head, one or the other migrating through and around the head during development. Some species face their "left" side upward, some face their "right" side upward, and others face either side upward. Some flatfish can camouflage themselves on the ocean floor.
- Wrasse are a large family of mainly small fish, usually less than 20 centimetres (7.9 in) long. Most wrasse are loners that prefer habitats such as coral reefs and rocky shores. They live close to the substrate, eating small invertebrates and almost anything else that lurks on the bottom. Many are brightly coloured. They have thick lips and use their sharp teeth to pick small creatures off the rocks. Many smaller wrasses follow the feeding trails of larger fish, picking up invertebrates disturbed by their passing.[:211]
- Triplefins are a family of fish. They are usually found around coral reefs and rocks, usually in shallow, clear sunlit waters such as lagoons and seaward reefs. Triplefins have three dorsal fins (hence the name). They are small fish, usually less than six cm long. Brightly coloured, often for reasons of camouflage, they are nervous and retreat to rock crevices at any perceived threat.
- Seahorses are a genus of fish. They prefer sheltered harbours, estuaries and other shallow coastal waters, where they hunt tiny crustaceans. They bob around in sheltered areas such as coral reefs, mangrove stands and seagrass meadows

and estuaries. They are camouflaged with murky patterns that blend into kelp and sea grass backgrounds. During social moments or in unusual surroundings, seahorses can turn on bright colours.

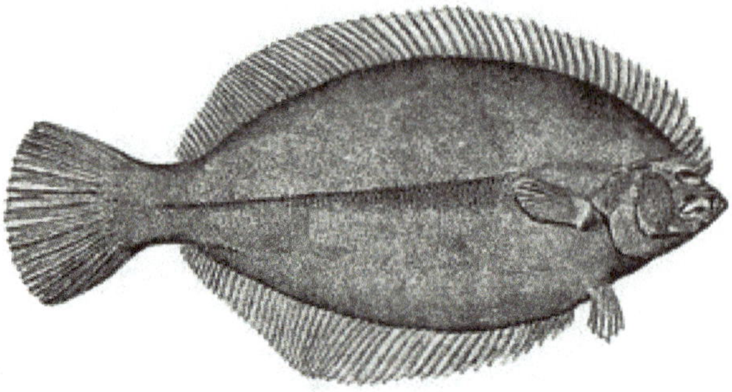

Flatfish

Cleaner wrasses working on a dragon wrasse

Common triplefin

Examples of shallow water pelagic fish, found in both tropical and temperate waters around the world, are grey mullet, sprats and garfish. As pelagic fish, all these fish spend most of their time living in the water column away the sea floor.

- The grey mullet are medium size fish, typically about 50 cm (20 in) long. They are often caught with set nets.
- The garfish is a long, slender fish, looking like a spear, which feeds on seagrass fragments, shrimps and crab larvae. In turn it is preyed on by larger fish and, since it is often near the surface, cormorants and gannets.

Grey mullet

Garfish

Plankton Feeding Pelagic Fish

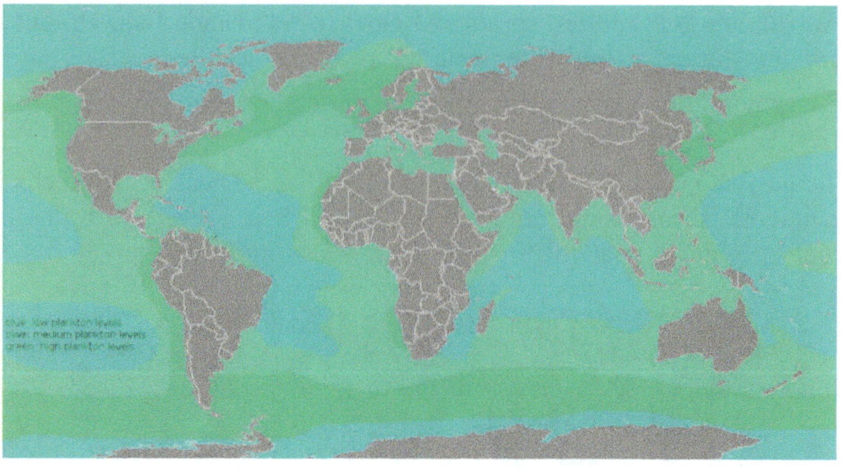

World distribution of plankton

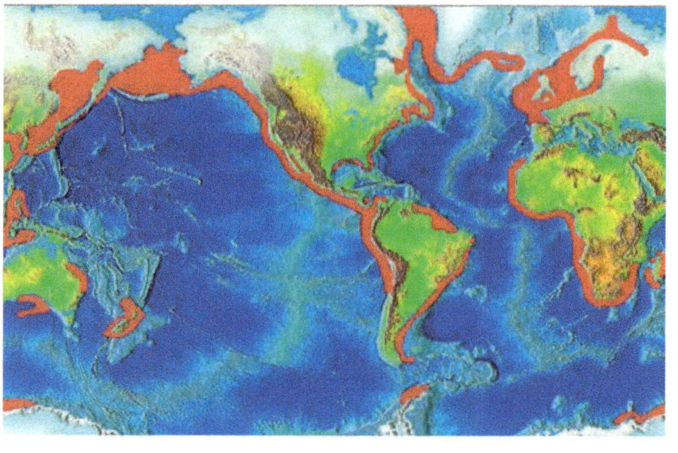

Areas of upwelling in red.

At the base of food chains are the primary producers. In the ocean these primary producers are mainly a type of plankton, microscopic phytoplankton which drift in the water column. Phytoplankton need sunlight for photosynthesis to power carbon fixation, so they are mainly located in sunlit surface waters. Phytoplankton also need and rapidly use nutrients in the water column. The phytoplankton are eaten by zooplankton, which in turn are eaten by predatory zooplankton. Filter feeders then eat the plankton and larger predatory fish eat the filter feeders (see diagram at the right).

Most filter-feeding pelagic fish found in coastal waters are small, silvery forage fish. Forage fish include fishes of the family Clupeidae (herrings, shad, sardines and pilchards, hilsa, menhaden and sprats), as well as anchovies, capelin and halfbeaks. They use schooling strategies to avoid predators, and different schools of forage fish often associate with each other in open coastal waters. Forage fish feed near the base of the food chain on plankton and fry (recently hatched fish), often by filter feeding. In turn, they are preyed on by larger predatory fish, seabirds and marine mammals.

Worldwide, there are five major coastal currents associated with upwelling areas: the Canary Current (off Northwest Africa), the Benguela Current (off southern Africa), the California Current (off California and Oregon), the Humboldt Current (off Peru and Chile), and the Somali Current (off Western India). All of these currents support major fisheries. Many forage fish are important commercial species, and the schools can be targeted by spotter planes. The fish are caught by purse seiners—fishing boats that use nets to enclose the fish—and can be overfished.

Schooling anchovies

European sprat

Atlantic herring

Underwater video loop of a school of herrings migrating to their spawning grounds in the Baltic Sea

Predatory Pelagic Fish

Predator bluefin trevally sizing up schoolinganchovies

Predatory pelagic fishes found on continental shelves worldwide in both tropical and temperate waters include porgies, barracuda, amberjacks and cutlassfishes. They tend to be larger fish, and are carnivorous, feeding on the smaller, silvery forage fish that eat plankton (see section above). Some species also feed on crabs and other invertebrates, foraged from the sea floor.

- Mackerel scom*
- Porgies sometimes called sea breams, are any of about 100 species belonging to

the family Sparidae. Porgies usually have high backs and a single dorsal fin, like snapper or grunt fishes (grunts are named for the sound they make grinding their teeth[:184]). They are bottom feeding pelagic fishes, with small mouths equipped with strong teeth adapted for handling small fishes and invertebrates with hard shells. Most do not exceed a size of about 30 cm (12 in), but some may grow to four times that length. They often school, and will migrate between reefs. Larger fish enter estuaries and harbours.

- Barracuda have long slender bodies typically about 50 cm (20 in) long. They have a wicked set of teeth and are ferocious predators. They feed on crustaceans, cephalopods and small fish like anchovy and pilchard. Barracouta often hunt in schools near the bottom or midwater, and sometimes even near the surface at night.

The jolthead porgy is a porgi grunt

Great barracuda and jacks

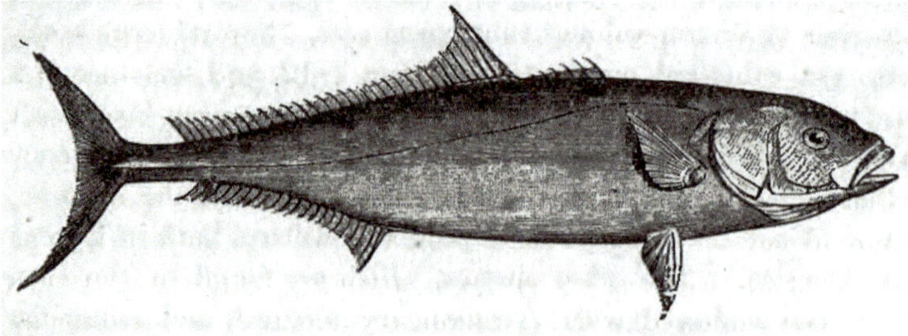

Yellow-tail (*Seriola lalandii*). (From South Australia.)

The yellowtail amberjack, pound for pound, is one of the hardest fighting fish in the ocean.

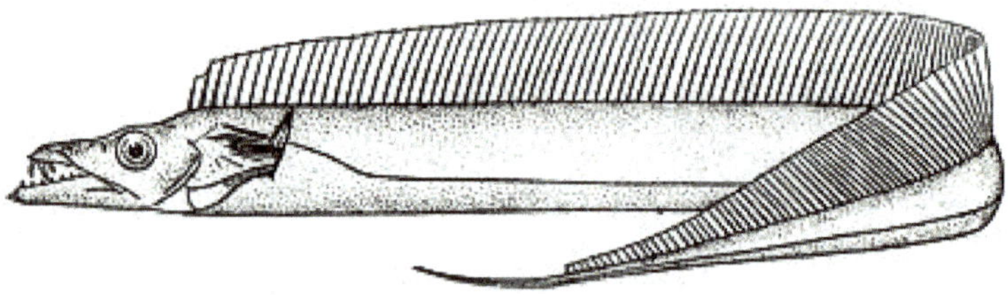

The largehead hairtail is a cutlassfish

The sand tiger shark is a large coastal shark that inhabits coastal waters worldwide. It's numbers are declining, and it is now listed as a vulnerable species on the IUCN Red List.

Giant trevally are great gamefish found in Indo-Pacific tropical waters. They are powerful apex predators in most of their habitats, hunting both individually and in schools.

- Cutlassfishes are a group of about 40 species belonging to the family Trichiuridae. They are ocean fish which regularly stray into coastal waters around the world. Fish of this family are long, slender, and generally steely blue or silver in colour, giving rise to their name. They have reduced or absent pelvic and caudal fins, giving them an eel-like appearance, and large fang-like teeth.[:190]
- Jacks, amberjacks, pompanos, horse mackerel, scads, leatherjackets and trevally are fish of the family *Carangidae*. Found in most coastal waters, they are fast predatory fishes that hunt in the waters above reefs and in the open sea; some dig in the sea floor for invertebrates (some can also filter feed, such as the white trevally). The largest fish in the family, the giant trevally, grows up to 1.7 m in length; most fish in the family reach a maximum length of 25–100 cm. The family contains many important commercial and game fish, notably the Pacific jack mackerel and the other jack mackerels in the genus *Trachurus*. The type species of this genus is the Atlantic horse mackerel. Jack mackerels are an important inshore commercial species.
- Amberjacks are a group of nine species belonging to the genus *Seriola* within the family Carangidae. Mainly open water fish, they can follow small forage fish into estuaries and enclosed waters, where they will also hunt for crustaceans. Amberjacks are fast swimming and aggressive predators that often hunt in schools around offshore reefs. the yellowtail amberjack can reach 1.8 metres in length and weigh 60 kilograms.

Demersal Fish

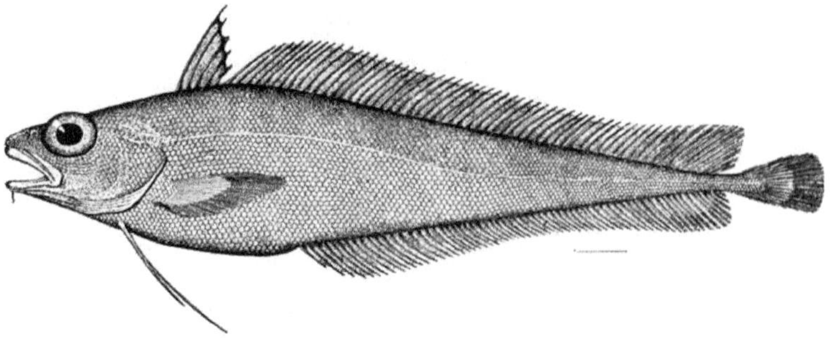

Cod-like fishes, like this morid cod have a barbel (fleshy filament) on their lower jaw which they use to detect prey buried in the sand or mud.

Fish that live on or in close association with the sea floor are called demersal fish. This section discusses the coastal demersal fish that live on the continental shelf, but are living further from the coast and in deeper water than the nearshore fish discussed above.

Demersal fish are white fish. Unlike oily fish, white fish contain oils only in their liver, rather than in the gut, and can therefore be gutted as soon as they are caught, on board the ship. White fish has dry and white flesh. They can be divided into benthopelagic fish (mostly round fish) which live *near* the sea bed, such as cod, and benthic fish (flatfish) such as plaice which live *on* the sea bed. Benthic fish tend to be "flat", so they can lie on the bottom.

- Cod-like fishes are a number of round benthopelagic species belonging to the order Gadiformes, such as Atlantic and Pacific cod, morid cod, haddock and pollock, including the highly commercial Alaska pollock. Cod-like fishes are often found in large schools over sandy or muddy bottoms. They have a barbel (fleshy filament) on their lower jaw which they use to detect prey buried in the sand or mud. Some migrate to warm water in winter to spawn.
- John dory are fishes of the genus *Zeus*. They have a widespread distribution and are typically found near the seabed in depths from 5 meters (16 feet). The John dory grows to a maximum length of 65 cm (2.13 ft). Although it is a benthopelagic fish, its body is flat and it can hardly be seen from the front because it is so thin. It is a poor swimmer with long spines on the dorsal fin. It has a large dark eyespot on the side of its body which is used to confuse prey, which are scooped up in its big mouth. Large eyes at the front of the head provide it with bifocal vision and depth perception, which are important for predators. The John Dory usually gets its food by stalking it then shooting out a tube in its mouth to capture its prey. It eats forage fish, and occasionally squid and cuttlefish. In turn, they are preyed on by sharks, like the dusky shark, and other large bony fish. They are normally solitary.
- Turbot and brill are benthic flatfish, resembling flounder and sole, but found

in deeper offshore waters on the continental shelf. They are brownish-green, with dark blotches on the turbot and mottling on the brill. They are fished by coastal trawlers.

The polar cod is found further north than any other fish species. It frequents river mouths and feeds on plankton and krill. It is preyed on by narwhals, belugas, ringed seals and seabirds.

The John dory is so thin it can hardly be seen from the front. The large eyespot on the side of its body confuses its prey.

The turbot is a large left-eyed flatfish usually found not too far from shore in sandy shallow waters. It is a prized food fish.

The red gurnard is a mail-cheeked fish found to depths of around 180 metres. They often grunt when captured as air is expelled from its air bladder.

- Mail-cheeked fishes belong to a group of about 30 species in the order Scorpaeniformes. Mail-cheeked fishes are named after a plate of bone that runs across each cheek. They are widespread in all the oceans of the world. Mail-cheeked fishes are carnivorous, mostly feeding on crustaceans, such as crabs and shrimp, and on smaller fish. Most species live on the sea bottom in relatively shallow waters, although species are known from mid and deep water, from the mid-water, and even from fresh water. They typically have spiny heads, and rounded pectoral and caudal fins. Most species are less than 30 centimetres (12 in) in length, but the full size range of the order varies from the velvetfishes, which can be just 2 centimetres (0.79 in) long as adults, to the Lingcod, which can reach 150 centimetres (4.9 ft) in length.

- Red gurnard are mail-cheeked fish. They use their large pectoral fins to rest on the bottom and to detect food.

The stargazer is an ambush predator which can deliver both venom and electric shocks. It has been called "the meanest thing in creation".

Stargazers are about 50 species of fishes, belonging to the family Uranoscopidae, and found worldwide in shallow waters. Stargazers are venomous; they have two large poison spines situated behind the opercle and above the pectoral fins. They can also deliver electric shocks. They are ambush predators with eyes on top of their heads (thus the name). Stargazers also have a large upward-facing mouth in a large head. They bury themselves in sand with only their eyes showing, and leap upwards to ambush fish and invertebrates overhead. Some species have a worm-shaped lure growing out of the floor of the mouth, which they wiggle to attract prey's attention. Lengths range from 18 cm up to 90 cm, for the giant stargazer *Kathetostoma giganteum*. Stargazers are a delicacy in some cultures. The venom is destroyed when it is cooked, and stargazers are sold in some fish markets with their electric organ removed. They have been called "the meanest things in creation" and the "worst pet on earth".

- Sandperches are a family, Pinguipedidae, containing 63 species of fishes in the order Perciformes. They are benthic carnivoures, feeding on small fish and invertebrates. Examples are the redbanded weever, yellow weaver and blue cod. They are often caught in pots like crayfish.
- Medusa fishes are a family Centrolophidae of 31 species of perciform fishes. They are found in temperate and tropical waters throughout the world, usually feeding on fish, crustaceans and small squid near rough sea floors on continental

shelf and slope. Examples are barrelfish, southern driftfish, imperial blackfish, the Japanese and pelagic butterfish, the New Zealand and Tasmanian ruffe, and the common, silver and white warehou. The young of some species associate with jellyfish, which provides them with protection from predators and opportunities to scavenge the remains of the jellyfish›s meals. The young of other species associate with large masses of floating kelp.

A speckled sand perch perched on coral sand

The rudderfish (*Centrolophus niger*) is a medusa fish

Malabar grouper

Atlantic wreckfish inhabit caves and wrecks. They are good game fish, reaching a maximum reported length of 210 cm (6.9 ft) and weight of 100 kg (220 lb).

- Grouper are fish belonging to a number of genera in the subfamily Epinephelinae of the family Serranidae, in the order Perciformes. Species of grouper include the black, comet, gag, giant, Goliath, Nassau, saddletail, tiger, Warsaw, white and yellowfin grouper. Typical lengths are 80–120 centimetres. They inhabit depths from reefs near the surface down to over 400 metres. They feed on just about any moving animal they encounter. Groper are important inshore commercial fish, usually caught with gill nets (in earlier times longlines were used).

- Wreckfish are a family Polyprionidae of perciform fishes, found on the floor of the continental shelf and slope where they inhabit caves and shipwrecks (thus their common name). The Atlantic wreckfish is at depths between 40 and 600 m (130 and 1,970 ft). They are largely a solitary fish, though juveniles school below floating objects. Their diet includes large ocean cephalopods, crustaceans, and other bottom-dwelling fishes.

REFERENCES

1. Spalding MD, Grenfell AM (1997). "New estimates of global and regional coral reef areas". Coral Reefs. 16 (4): 225–230. doi:10.1007/s003380050078.
2. Spalding, Mark, Corinna Ravilious, and Edmund Green (2001). World Atlas of Coral Reefs. Berkeley, CA: University of California Press and UNEP/WCMC ISBN 0520232550.
3. Mulhall M (Spring 2009) Saving rainforests of the sea: An analysis of international efforts to conserve coral reefs Duke Environmental Law and Policy Forum 19:321–351.
4. Where are Corals Found? NOAA. Reviewed: 13 May 2011. Retrieved: 24 March 2015.
5. Hoover, John (November 2007). Hawai□i's Sea Creatures. Mutual. ISBN 1-56647-220-2.
6. Cesar, H.J.S; Burke, L.; Pet-Soede, L. (2003). The Economics of Worldwide Coral Reef Degradation (PDF). The Netherlands: Cesar Environmental Economics Consulting. p. 4. Retrieved 21 September 2013.
7. Costanza, Robert; Ralph d'Arge; Rudolf de Groot; Stephen Farber; Monica Grasso; Bruce Hannon; Karin Limburg; Shahid Naeem; Robert V. O'Neill; Jose Paruelo; Robert G. Raskin; Paul Sutton; Marjan van den Belt (15 May 1997). "The value of the world's ecosystem services and natural capital". Nature. 387 (6630): 253–260. Bibcode:1997Natur.387..253C. doi:10.1038/387253a0.
8. Danovaro, Roberto; Bongiorni, Lucia; Corinaldesi, Cinzia; Giovannelli, Donato; Damiani, Elisabetta; Astolfi, Paola; Greci, Lucedio; Pusceddu, Antonio (April 2008). "Sunscreens Cause Coral Bleaching by Promoting Viral Infections". Environmental Health Perspectives. 116 (4): 441–447. doi:10.1289/ehp.10966. PMC 2291018 . PMID 18414624.
9. "Corals reveal impact of land use". ARC Centre of Excellence for Coral Reef Studies. Retrieved September 21, 2013.
10. Minato, Charissa (July 1, 2002). "Urban runoff and coastal water quality being researched for effects on coral reefs" (PDF). Retrieved December 2010. Check date values in: |access-date= (help)
11. "Coastal Watershed Factsheets – Coral Reefs and Your Coastal Watershed". Environmental Protection Agency Office of Water. July 1998. Retrieved December 2010. Check date values in: |access-date= (help)
12. Kleypas, Joanie (September 21, 2010). "Coral reef". The Encyclopedia of Earth.

Retrieved April 4, 2011.

13. Darwin, Charles (1842). “The Structure and Distribution of Coral Reefs. Being the first part of the geology of the voyage of the Beagle, under the command of Capt. Fitzroy, R.N. during the years 1832 to 1836”. London: Smith Elder and Co.
14. Chancellor, Gordon (2008). “Introduction to Coral reefs”. Darwin Online. Retrieved January 20, 2009.
15. Animation of coral atoll formation NOAA Ocean Education Service. Retrieved January 9, 2010.
16. Webster, Jody M.; Braga, Juan Carlos; Clague, David A.; Gallup, Christina; Hein, James R.; Potts, Donald C.; Renema, Willem; Riding, Robert; Riker-Coleman, Kristin; Silver, Eli; Wallace, Laura M. (1 March 2009). “Coral reef evolution on rapidly subsiding margins”. Global and Planetary Change. 66 (1–2): 129–148. Bibcode:2009GPC....66..129W. doi:10.1016/j.gloplacha.2008.07.010.
17. Webster, Jody M.; Clague, David A.; Riker-Coleman, Kristin; Gallup, Christina; Braga, Juan C.; Potts, Donald; Moore, James G.; Winterer, Edward L.; Paull, Charles K. (1 January 2004). “Drowning of the −150 m reef off Hawaii: A casualty of global meltwater pulse 1A?”. Geology. 32 (3): 249. Bibcode:2004Geo....32..249W. doi:10.1130/G20170.1.
18. Great Barrier Reef Marine Park Authority (2006). “A “big picture” view of the Great Barrier Reef” (PDF). Reef Facts for Tour Guides. Retrieved June 18, 2007.
19. Tobin, Barry (2003) . “How the Great Barrier Reef was formed”. Australian Institute of Marine Science. Retrieved November 22, 2006.
20. CRC Reef Research Centre Ltd. “What is the Great Barrier Reef?”. Archived from the original on August 22, 2006. Retrieved May 28, 2006.
21. Four Types of Coral Reef Microdocs, Stanford Education. Retrieved January 10, 2010.
22. MSN Encarta (2006). Great Barrier Reef. Archived from the original on October 31, 2009. Retrieved December 11, 2006.
23. Murphy, Richard C. (2002). Coral Reefs: Cities Under The Seas. The Darwin Press, Inc. ISBN 0-87850-138-X.
24. Smithers, S.G.; Woodroffe, C.D. (2000). “Microatolls as sea-level indicators on a mid-ocean atoll”. Marine Geology. 168 (1–4): 61–78. doi:10.1016/S0025-3227(00)00043-8.
25. Moyle, Peter B.; Joseph J. Cech (2004). Fishes : an introduction to ichthyology (Fifth ed.). Upper Saddle River, N.J.: Pearson/Prentice Hall. p. 556. ISBN 978-0-13-100847-2.
26. Connell, Joseph H. (March 24, 1978). “Diversity in Tropical Rain Forests and Coral Reefs”. Science. 199 (4335): 1302–1310. Bibcode:1978Sci...199.1302C. doi:10.1126/science.199.4335.1302. PMID 17840770.
27. UNEP (2001) UNEP-WCMC World Atlas of Coral Reefs Coral Reef Unit
28. Achituv, Y. and Dubinsky, Z. 1990. Evolution and Zoogeography of Coral Reefs

Ecosystems of the World. Vol. 25:1–8.

29. Wells, Sue; Hanna, Nick (1992). Greenpeace Book of Coral Reefs. Sterling Publishing Company. ISBN 0-8069-8795-2.
30. Vajed Samiei, J.; Dab K.; Ghezellou P.; Shirvani A. (2013). "Some Scleractinian Corals (Class: Anthozoa) of Larak Island, Persian Gulf". Zootaxa. 3636 (1): 101–143. doi:10.11646/zootaxa.3636.1.5.
31. Gunnerus, Johan Ernst (1768). Om Nogle Norske Coraller.
32. Nybakken, James. 1997. Marine Biology: An Ecological Approach. 4th ed. Menlo Park, CA: Addison Wesley.
33. NOAA CoRIS – Regional Portal – Florida. Coris.noaa.gov (August 16, 2012). Retrieved on March 3, 2013.
34. NGM.nationalgeographic.com, Ultra Marine: In far eastern Indonesia, the Raja Ampat islands embrace a phenomenal coral wilderness, by David Doubilet, National Geographic, September 2007
35. Living Reefs Foundation. Retrieved on May 28, 2015.
36. LiveScience. Retrieved on April 14, 2016.
37. Sherman, C.D.H. "The Importance of Fine-scale Environmental Heterogeneity in Determining Levels of Genotypic Diversity and Local Adaption." University of Wollongong Ph.D. Thesis. 2006. Accessed June 7, 2009.
38. Marshall, Paul; Schuttenberg, Heidi (2006). A Reef Manager's Guide to Coral Bleaching. Townsville, Australia: Great Barrier Reef Marine Park Authority. ISBN 1-876945-40-0.
39. Stacy, J., Marion, G., McCulloch, M. and Hoegh-Guldberg, O. "changes to Mackay Whitsunday water quality and connectivity between terrestrial, mangrove and coral reef ecosystems: Clues from coral proxies and remote sensing records -Synthesis of research from an ARC Linkage Grant (2004–2007)." University of Queensland – Centre for Marine Studies. May 2007. Accessed June 7, 2009.
40. Nothdurft, L.D. "Microstructure and early diagenesis of recent reef building scleractinian corals, Heron Reef, Great Barrier Reef: Implications for palaeoclimate analysis." Queensland University of Technology Ph.D. Thesis. 2007. Accessed June 7, 2009.
41. Wilson, R.A. "The Biological Notion of Individual."Stanford Encyclopedia of Philosophy. August 9, 2007. Accessed June 7, 2009.
42. Jennings S, Kaiser MJ and Reynolds JD (2001) Marine fisheries ecology, Wiley-Blackwell, pp. 291–293. ISBN 978-0-632-05098-7.
43. Chappell, John (17 July 1980). "Coral morphology, diversity and reef growth". Nature. 286 (5770): 249–252. Bibcode:1980Natur.286..249C. doi:10.1038/286249a0.
44. Jackson, Jeremy B. C. (1 July 1991). "Adaptation and Diversity of Reef Corals". BioScience. 41 (7): 475–482. doi:10.2307/1311805. JSTOR 1311805.
45. Reynolds, J., B. Bruns, W. Fitt, G. Schmidt. 2008. Enhanced photoprotection

pathways in symbiotic dinoflagellates of shallow-water corals and other cnidarians. 13674–13678, doi:10.1073/pnas.0805187105

46. Rougerier, F (1998). "The functioning of coral reefs and atolls: from paradox to paradigm". In Jost, Christian. The French-Speaking Pacific: Population, Environment and Development Issues (PDF). Boombana Publications. ISBN 978-1-876542-02-3.
47. Darwin, Charles R. (1842). The Structure and Distribution of Coral Reefs. Being the first part of the geology of the voyage of the Beagle, under the command of Capt. Fitzroy, R.N. during the years 1832 to 1836. London: Smith Elder and Co. pp. 61–71.
48. Crossland CJ (1983) "Dissolved nutrients in coral reef waters In DJ Barnes (Ed) Perspectives on coral reefs, pages 56–68, Australian Institute of Marine Science. ISBN 9780642895851.
49. Odum EP (1971) Fundamentals of ecology Saunders.
50. Sammarco, PW; Risk, MJ; Schwarcz, HP; Heikoop, JM (1999). "Cross-continental shelf trends in coral δ15N on the Great Barrier Reef: further consideration of the reef nutrient paradox" (PDF). Mar Ecol Prog Ser. 180: 131–138. doi:10.3354/meps180131.
51. Rougerie, F; Wauthy, B (1993). "The endo-upwelling concept: from geothermal convection to reef construction" (PDF). Coral Reefs. 12: 19–30. Bibcode:1993CorRe..12...19R. doi:10.1007/bf00303781.
52. De Goeij, Jasper M (2009) "Element cycling on tropical coral reefs: the cryptic carbon shunt revealed" PhD thesis, page 13. University of Groningen.
53. Sorokin, Yuri I. (1993). Coral Reef Ecology. Germany: Springer-Verlag, Berlin Heidelberg. ISBN 978-0-387-56427-2.
54. Hatcher, Bruce Gordon (1 May 1988). "Coral reef primary productivity: A beggar's banquet". Trends in Ecology & Evolution. 3 (5): 106–111. doi:10.1016/0169-5347(88)90117-6.
55. Castro, Peter and Huber, Michael (2000) Marine Biology. 3rd ed. Boston: McGraw-Hill.
56. Zooxanthellae… What's That?. Oceanservice.noaa.gov (March 25, 2008). Retrieved on November 1, 2011.
57. Roach, John (November 7, 2001). "Rich Coral Reefs in Nutrient-Poor Water: Paradox Explained?". National Geographic News. Retrieved April 5, 2011.
58. Nowak (21 September 2002). "Corals play rough over Darwin's paradox". New Scientist (2361). |first1= missing |last1= in Authors list (help)
59. Leichter, J.; Wing S.; Miller S.; Denny M. (1996). "Pulsed delivery of subthermocline water to Conch Reef (Florida Keys) by internal tidal bores". Limnology and Oceanography. 41 (7): 1490–1501. doi:10.4319/lo.1996.41.7.1490.
60. Wolanski, E.; Pickard, G. L. (1983). "Upwelling by internal tides and kelvin waves at the continental shelf break on the Great Barrier Reef". Marine and Freshwater Research. 34: 65. doi:10.1071/MF9830065.

61. Leichter, J.; Helmuth B.; Fischer A. (2006). "Variation beneath the surface: Quantifying complex thermal environments on coral reefs in the Caribbean, Bahamas and Florida". Journal of Marine Research. 64 (4): 563–588. doi:10.1357/002224006778715711.

62. Ezer, T.; Heyman W.; Houser C.; Kjerfve B. (2011). "Modeling and observations of high-frequency flow variability and internal waves at a Caribbean reef spawning aggregation site". Ocean Dynamics. 61 (5): 581–598. Bibcode:2011OcDyn..61..581E. doi:10.1007/s10236-010-0367-2.

63. Fratantoni, D.; Richardson P. (2006). "The Evolution and Demise of North Brazil Current Rings". Journal of Physical Oceanography. 36 (7): 1241–1249. Bibcode:2006JPO....36.1241F. doi:10.1175/JPO2907.1.

64. Leichter, J.; Shellenbarger G.; Genovese S.; Wing S. (1998). "Breaking internal waves on a Florida (USA) coral reef: a plankton pump at work?". Marine Ecology Progress Series. 166: 83–97. doi:10.3354/meps166083.

65. Talley, L. (2011). Descriptive Physical Oceanography: An Introduction. Oxford UK: Elsevier Inc. ISBN 0750645520.

66. Helfrich, K. (1992). "Internal solitary wave breaking and run-up on a uniform slope". Journal of Fluid Mechanics. 243: 133–154. Bibcode:1992JFM...243..133H. doi:10.1017/S0022112092002660.

67. Gregg, M. (1989). "Scaling turbulent dissipation in the thermocline". Journal of Geophysical Research. 9686–9698. 94: 9686. Bibcode:1989JGR....94.9686G. doi:10.1029/JC094iC07p09686.

68. Taylor, J. (1992). "The energetics of breaking events in a resonantly forced internal wave field". Journal of Fluid Mechanics. 239: 309– 340. Bibcode:1992JFM...239..309T. doi:10.1017/S0022112092004427.

69. Andrews, J.; Gentien P. (1982). "Upwelling as a source of nutrients for the Great Barrier Reef ecosystems: A solution to Darwin's question?". Marine Ecology Progress Series. 8: 257–269. doi:10.3354/meps008257.

70. Sandstrom, H.; Elliott J. (1984). "Internal tide and solitons on the Scotian shelf: A nutrient pump at work". Journal of Geophysical Research. 89: 6415–6426. Bibcode:1984JGR....89.6415S. doi:10.1029/JC089iC04p06415.

71. Wolanski, E.; Hamner W. (1988). "Topographically controlled fronts in the ocean and their biological significance". Science. 241 (4862): 177–181. Bibcode:1988Sci...241..177W. doi:10.1126/science.241.4862.177. PMID 17841048.

72. Rougerie, F.; Fagerstrom J.; Andrie C. (1992). "Geothermal endo-upwelling: A solution to the reef nutrient paradox?". Continental Shelf Research. 12 (7–8): 785–798. Bibcode:1992CSR....12..785R. doi:10.1016/0278-4343(92)90044-K.

73. Wolanski, E.; Delesalle B. (1993). "Upwelling by internal waves, Tahiti, French Polynesia". Continental Shelf Research. 15 (2–3): 357–368. Bibcode:1995CSR....15..357W. doi:10.1016/0278-4343(93)E0004-R.

74. Szmant, A. M.; Forrester, A. (1996). "Water column and sediment nitrogen and

phosphorus distribution patterns in the Florida Keys, USA". Coral Reefs. 15: 21–41. Bibcode:1996CorRe..15...21S. doi:10.1007/BF01626075.

75. Furnas, M. J.; Mitchell, A. W. (1996). "Nutrient inputs into the central Great Barrier Reef (Australia) from subsurface intrusions of Coral Sea waters: A two-dimensional displacement model". Continental Shelf Research. 16 (9): 1127–1148. Bibcode:1996CSR....16.1127F. doi:10.1016/0278-4343(95)00060-7.

76. Leichter, J.; Miller S. (1999). "Predicting high-frequency upwelling: Spatial and temporal patterns of temperature anomalies on a Florida coral reef.". Continental Shelf Research. 19: 911–928. doi:10.1016/s0278-4343(99)00004-7.

77. Leichter, J.; Stewart H.; Miller S. (2003). "Episodic nutrient transport to Florida coral reefs". Limnology and Oceanography. 48 (4): 1394–1407. doi:10.4319/lo.2003.48.4.1394.

78. Leichter, J.; Deane G.; Stokes M. (2005). "Spatial and Temporal Variability of Internal Wave Forcing on a Coral Reef". Journal of Physical Oceanography. 35 (11): 1945–1962. Bibcode:2005JPO....35.1945L. doi:10.1175/JPO2808.1.

79. Smith, J.; Smith C.; Vroom P.; Beach K.; Miller S. (2004). "Nutrient and growth dynamics of Halimeda tuna on Conch Reef, Florida Keys: Possible influence of internal tides on nutrient status and physiology". Limnology and Oceanography. 49 (6): 1923–1936. doi:10.4319/lo.2004.49.6.1923.

80. Pineda, J. (1994). "Internal tidal bores in the nearshore: Warm-water fronts, seaward gravity currents and the onshore transport of neustonic larvae". Journal of Marine Research. 52 (3): 427–458. doi:10.1357/0022240943077046.

81. Wilson, E (2004). "Coral's Symbiotic Bacteria Fluoresce, Fix Nitrogen". Chemical and Engineering News. 82 (33): 7.

82. Barnes, R.S.K.; Mann, K.H. (1991). Fundamentals of Aquatic Ecology. Blackwell Publishing. pp. 217–227. ISBN 0-632-02983-8.

83. Fuchs. T (2013). "Effects of Coral Reef Complexity on Invertebrate Biodiversity". Immediate Science Ecology Publishing: 1–10.

84. Hatcher, B.G. Johannes, R.E.; Robertson, A.J. (1989). "Conservation of Shallow-water Marine Ecosystems". Oceanography and Marine Biology: An Annual Review. 27. Routledge. p. 320. ISBN 0-08-037718-1.

85. "World's Reef Fishes Tussling With Human Overpopulation". ScienceDaily. April 5, 2011.

86. "Coral Reef Biology". NOAA. Retrieved April 6, 2011.

87. Glynn, P.W. (1990). Dubinsky, Z., ed. Ecosystems of the World v. 25-Coral Reefs. New York, NY: Elsevier Science. ISBN 978-0-444-87392-7.

88. "THE EFFECTS OF TERRESTRIAL RUNOFF OF SEDIMENTS, NUTRIENTS AND OTHER POLLUTANTS ON CORAL REEFS" (PDF). Retrieved 2015-12-05.

89. Vroom, Peter S.; Page, Kimberly N.; Kenyon, Jean C.; Brainard, Russell E. (2006). "Algae-Dominated Reefs". American Scientist. 94 (5): 430–437. doi:10.1511/2006.61.1004.

90. Kaplan, Matt (2009). "How the sponge stays slim". Nature. doi:10.1038/news.2009.1088.
91. Buchheim, Jason. "Coral Reef Fish Ecology". marinebiology.org. Retrieved April 5, 2011.
92. McClellan, Kate; Bruno, John (2008). "Coral degradation through destructive fishing practices". Encyclopedia of Earth. Retrieved October 25, 2008.
93. Osborne, Patrick L. (2000). Tropical Ecosystem and Ecological Concepts. Cambridge: Cambridge University Press. p. 464. ISBN 0-521-64523-9.
94. The.honoluluadvertiser.com. The.honoluluadvertiser.com (January 17, 2005). Retrieved on November 1, 2011.
95. "U.S. Fish & Wildlife Service – Birds of Midway Atoll". Retrieved August 19, 2009.
96. Heatwole, Harold (1999). Sea snakes (2. ed.). Malabar, Fla: Krieger. ISBN 1-57524-116-1.
97. Li, Min; Fry, B.G.; Kini, R. Manjunatha (1 January 2005). "Eggs-Only Diet: Its Implications for the Toxin Profile Changes and Ecology of the Marbled Sea Snake (Aipysurus eydouxii)". Journal of Molecular Evolution. 60 (1): 81–89. doi:10.1007/s00239-004-0138-0. PMID 15696370.
98. Voris, Harold K. (1 January 1966). "Fish Eggs as the Apparent Sole Food Item for a Genus of Sea Snake, Emydocephalus (Krefft)". Ecology. 47 (1): 152. doi:10.2307/1935755.
99. McClenachan, Loren; Jackson, Jeremy BC; Newman, Marah JH (1 August 2006). "Conservation implications of historic sea turtle nesting beach loss". Frontiers in Ecology and the Environment. 4 (6): 290–296. doi:10.1890/1540-9295(2006)4290:CIOHST2.0.CO;2.
100. Lutz, Peter L.; Musick, John A. (1996). The biology of sea turtles. Boca Raton, Fla: CRC Press. ISBN 0849384222.
101. Meylan, Anne (January 22, 1988). "Spongivory in Hawksbill Turtles: A Diet of Glass". Science. 239 (4838): 393–395. Bibcode:1988Sci...239..393M. doi:10.1126/science.239.4838.393. PMID 17836872.
102. "The Importance of Coral to People". World Wildlife Fund. Retrieved April 7, 2011.
103. "Coastal Capital: Economic Valuation of Coastal Ecosystems in the Caribbean". World Resources Institute.
104. Cooper, Emily; Burke, Lauretta; Bood, Nadia (2008). "Coastal Capital: Belize: The Economic Contribution of Belize's Coral Reefs and Mangroves" (PDF). Retrieved April 6, 2011.
105. Sarkis, Samia; van Beukering, Pieter J.H.; McKenzie, Emily (2010). "Total Economic Value of Bermuda's Coral Reefs. Valuation of ecosystem Services"(PDF). Retrieved May 29, 2015.
106. "Coral reefs around the world". Guardian.co.uk. September 2, 2009.

107. "Caribbean coral reefs may disappear within 20 years: Report". IANS. news. biharprabha.com. Retrieved 3 July 2014.
108. Liz Minchin (8 April 2013). "Air pollution casts a cloud over coral reef growth". The Conversation. The Conversation Media Group. Retrieved 9 April 2013.
109. Clapham ME and Payne (2011). "Acidification, anoxia, and extinction: A multiple logistic regression analysis of extinction selectivity during the Middle and Late Permian". Geology. 39 (11): 1059–1062. Bibcode:2011Geo....39.1059C. doi:10.1130/G32230.1.
110. Payne JL, Clapham ME (2012). "End-Permian Mass Extinction in the Oceans: An Ancient Analog for the Twenty-First Century?". Annual Review of Earth and Planetary Sciences. 40: 89–111. Bibcode:2012AREPS..40...89P. doi:10.1146/annurev-earth-042711-105329.
111. Life in the Sea Found Its Fate in a Paroxysm of Extinction New York Times, April 30, 2012.
112. Hutchings, P.A. (1986). "Biological destruction of coral reefs". Coral Reefs. Springer-Verlag. 12 (1): 1–17. Bibcode:1986CorRe...4..239H. doi:10.1007/BF00298083.
113. Losing Our Coral Reefs – Eco Matters – State of the Planet. Blogs.ei.columbia.edu. Retrieved on November 1, 2011.
114. Ritter, Karl (December 8, 2010). −goal-coral-reefs.html "Climate goal may spell end for some coral reefs". Associated Press. Retrieved December 2010. Check date values in: |access-date= (help)
115. Markey, Sean (May 16, 2006). "Global Warming Has Devastating Effect on Coral Reefs, Study Shows". National Geographic News.
116. Abrego D, Ulstrup KE, Willis BL, van Oppen MJH (2008) Species–specific interactions between algal endosymbionts and coral hosts define their bleaching response to heat and light stress. Proc Biol Sci 275: 2273–2282
117. Kleypas, J.A.; Feely, R.A.; Fabry, V.J.; Langdon, C.; Sabine, C.L.; Robbins, L.L. (2006). "Impacts of Ocean Acidification on Coral Reefs and Other Marine Calcifiers: A guide for Future Research" (PDF). National Science Foundation, NOAA, & United States Geological Survey. Retrieved April 7, 2011.
118. Tun, K.; Chou, L.M.; Cabanban, A.; Tuan, V.S.; Philreefs; Yeemin, T.; Suharsono; Sour, K.; Lane, D. (2004). "Status of Coral Reefs, Coral Reef Monitoring and Management in Southeast Asia, 2004". In Wilkinson, C. Status of Coral Reefs of the world: 2004 (PDF). Townsville, Queensland, Australia: Australian Institute of Marine Science. pp. 235–276.
119. "Reefs at Risk Revisited: Southeast Asia" (PDF). World Resources Institute. February 2011. Retrieved January 27, 2014.
120. "Reefs at Risk Revisited". World Resources Institute. February 2011. pp. 45–46. Retrieved January 27, 2014.
121. Lamb, Joleah; Bette Willis (August 16, 2011). "Using coral disease prevalence to assess the effects of concentrating tourism activities on offshore reefs in a

tropical marine park". Conservation Biology. 25 (5): 1044–1052. doi:10.1111/j.1523-1739.2011.01724.x. PMID 21848962.

122. McClanahan, Timothy; Marnane, Michael; Cinner, Joshua E.; Kiene, William E. (2006). "A Comparison of Marine Protected Areas and Alternative Approaches to Coral-Reef Management". Current Biology. 16 (14): 1408–13. doi:10.1016/j.cub.2006.05.062. PMID 16860739.

123. Christie, P. (2004). "Marine protected areas as biological successes and social failures in Southeast Asia". American Fisheries Society Symposium. 2004 (42): 155–164.

124. McClanahan, Timothy; Davies, Jamie; Maina, Joseph (2005). "Factors influencing resource users and managers' perceptions towards marine protected area management in Kenya". Environmental Conservation. 32: 42–49. doi:10.1017/S0376892904001791.

125. Stone, Gregory (January 2011). "Phoenix Rising". National Geographic Magazine.

126. Ewa Magiera; Sylvie Rockel (2 July 2014). "From despair to repair: Dramatic decline of Caribbean corals can be reversed". Retrieved 8 June 2015.

127. Kelly, R.P; Foley; Fisher, WS; Feely, RA; Halpern, BS; Waldbusser, GG; Caldwell, MR; et al. (2011). "Mitigating local causes of ocean acidification with existing laws" (PDF). Science. 332 (6033): 1036–1037. Bibcode:2011Sci...332.1036K. doi:10.1126/science.1203815. PMID 21617060.

128. "Satellites to assess coral reef health". Retrieved 13 December 2014.

129. "World Heritage List". UNESCO. Retrieved 18 December 2016.

130. "A biodiversity strategy for the Great Barrier Reef". Great Barrier Reef Marine Park Authority, Australian Government. Retrieved 20 September 2013.

131. "Great Barrier Reef Climate Change Action Plan 2007–2011" (PDF). Great Barrier Reef Marine Park Authority. 2007.

132. Cinner, Joshua E.; Marnane, Michael J.; McClanahan, Tim R. (2005). "Conservation and community benefits from traditional coral reef management at Ahus Island, Papua New Guinea". Conservation Biology. 19 (6): 1714–1723. doi:10.1111/j.1523-1739.2005.00209.x-i1.

133. "Coral Reef Management, Papua New Guinea". Nasa's Earth Observatory. Retrieved November 2, 2006.

134. Horoszowski-Fridman, YB, Izhaki, I & Rinkevich, B; Izhaki; Rinkevich (2011). "Engineering of coral reef larval supply through transplantation of nursery-farmed gravid colonies". Journal of Experimental Marine Biology and Ecology. 399 (2): 162–166. doi:10.1016/j.jembe.2011.01.005.

135. Pomeroy, RS, Parks, JE and Balboa, CM (2006). "Farming the reef: is aquaculture a solution for reducing fishing pressure on coral reefs?". Marine Policy. 30 (2): 111–130. doi:10.1016/j.marpol.2004.09.001.

136. Rinkevich, B (2008). "Management of coral reefs: We have gone wrong when neglecting active reef restoration" (PDF). Marine pollution bulletin. 56 (11):

1821–1824. doi:10.1016/j.marpolbul.2008.08.014. PMID 18829052.

137. Ferse, SCA (2010). "Poor Performance of Corals Transplanted onto Substrates of Short Durability". Restoration Ecology. 18 (4): 399–407. doi:10.1111/j.1526-100X.2010.00682.x.

138. Superglue used for placement of coral. coralgarden.co.uk (May 8, 2009). Retrieved on November 8, 2011.

139. Needle and thread use with soft coral. coralgarden.co.uk (May 8, 2009). Retrieved on November 8, 2011.

140. Sabater, Marlowe G.; Yap, Helen T. (2004). "Long-term effects of induced mineral accretion on growth, survival, and corallite properties of Porites cylindrica Dana". Journal of Experimental Marine Biology and Ecology. 311 (2): 355–374. doi:10.1016/j.jembe.2004.05.013.

141. Jokeil, P.L.; Ku'lei, S.R (2004). "Coral Relocation Project in Kaneohe Bay, Oahu, Hawaii: Report on Phase 1" (PDF). University of Hawaii.

142. "Gene Therapy Could Help Corals Survive Climate Change". Scientific American. February 29, 2012.

143. Cobb, K.; Charles, Christopher D.; Cheng, Hai; Edwards, R. Lawrence (2003). "El Nino/Southern Oscillation and tropical Pacific climate during the past millennium". Nature. 424 (6946): 271–6. Bibcode:2003Natur.424..271C. doi:10.1038/nature01779. PMID 12867972.

CHAPTER 5

Pelagic Sediment

Chapter Contents

Pelagic sediment comprises fine-grained material composed of lithogenous, biogenous, and cosmogenous particles that have been deposited in the ocean floor away from the neritic zone. Pelagic sediment is composed of calcareous and siliceous shells from marine organisms, fine-grained clay material, volcanic ash and trace amounts of meteor debris. This material has been deposited, at slow depositional rates, in the ocean floor by wind, ice glaciers, rivers, and the skeletal remains of surface dwelling microorganisms.

INTRODUCTION

Pelagic sediment or pelagite is a fine-grained sediment that accumulates as the result of the settling of particles to the floor of the open ocean, far from land. These particles consist primarily of either the microscopic, calcareous or siliceous shells of phytoplankton or zooplankton; clay-size siliciclastic sediment; or some mixture of these. Trace amounts of meteoric dust and variable amounts of volcanic ash also occur within pelagic sediments. Based upon the composition of the ooze, there are three main types of pelagic sediments: siliceous oozes, calcareous oozes, and red clays.

The composition of pelagic sediments is controlled by three main factors. The first factor is the distance from major landmasses, which affects their dilution by terrigenous, or land-derived, sediment. The second factor is water depth, which affects the preservation of both siliceous and calcareous biogenic particles as they settle to the ocean bottom. The final factor is ocean fertility, which controls the amount of biogenic particles produced in surface waters.

OOZES

In case of marine sediments, ooze does not refer to a sediment's consistency, but to its composition, which directly reflects its origin. Ooze is pelagic sediment that consists of at least 30% of microscopic remains of either calcareous or siliceous planktonic debris organisms. The remainder typically consists almost entirely of clay minerals. As a result, the grain size of oozes is often bimodal with a well-defined biogenic silt- to sand-size fraction and siliciclastic clay-size fraction. Oozes can be defined by and classified according to the predominate organism that compose them. For example, there are diatom, coccolith, foraminifera, globigerina, pteropod, and radiolarian oozes. Oozes are also classified and named according to their mineralogy, i.e. calcareous or siliceous oozes. Whatever their composition, all oozes accumulate extremely slowly, at no more than a few centimeters per millennium.

Calcareous ooze is ooze that is composed of at least 30% of the calcareous microscopic shells—also known as *tests*—of foraminifera, coccolithophores, and pteropods. This is the most common pelagic sediment by area, covering 48% of the world ocean's floor. This type of ooze accumulates on the ocean floor at depths above the carbonate compensation depth. It accumulates more rapidly than any other pelagic sediment type, with a rate that varies from 0.3–5 cm/1000 yr.

Siliceous ooze is ooze that is composed of at least 30% of the siliceous microscopic "shells" of plankton, such as diatoms and radiolaria. Siliceous oozes often contain lesser proportions of either sponge spicules, silicoflagellates or both. This type of ooze accumulates on the ocean floor at depths below the carbonate compensation depth. Its distribution is also limited to areas with high biological productivity, such as the polar oceans, and upwelling zones near the equator. The least common type of sediment, it covers only 15% of the ocean floor. It accumulates at a slower rate than calcareous ooze: 0.2–1 cm/1000 yr.

RED AND BROWN CLAYS

Red clay, also known as either *brown clay* or *pelagic clay,* accumulates in the deepest and most remote areas of the ocean. It covers 38% of the ocean floor and accumulates more slowly than any other sediment type, at only 0.1-0.5 cm/1000 yr. Containing less than 30% biogenic material, it consists of sediment that remains after the dissolution of both calcareous and siliceous biogenic particles while they settled through the water column. These sediments consist of aeolianquartz, clay minerals, volcanic ash, subordinate residue of siliceous microfossils, and authigenic minerals such as zeolites, limonite and manganese oxides. The bulk of red clay consists of eolian dust. Accessory constituents found in red clay include meteorite dust, fish bones and teeth, whale ear bones, and manganese micro-nodules.

These pelagic sediments are typically bright red to chocolate brown in color. The color results from coatings of iron and manganese oxide on the sediment particles. In the absence of organic carbon, iron and manganese remain in their oxidized states and these clays remain brown after burial. When more deeply buried, brown clay may change into red clay due to the conversion of iron-hydroxides to hematite.

These sediments accumulate on the ocean floor within areas characterized by little planktonic production. The clays which comprise them were transported into the deep ocean in suspension, either in the air over the oceans or in surface waters. Both wind and ocean currents transported these sediments in suspension thousands of kilometers from their terrestrial source. As they were transported, the finer clays may have stayed in suspension for a hundred years or more within the water column before they settled to the ocean bottom. The settling of this clay-size sediment occurred primarily by the formation of clay aggregates by flocculation and by their incorporation into fecal pellets by pelagic organisms.

Distribution and Average Thickness of Marine Sediments

Region	Percent of Ocean Area	Percent of Total Volume of Marine Sediments	Average Thickness
Continental shelves	9%	15%	2.5 km (1.6 mi)
Continental slopes	6%	41%	9 km (5.6 mi)
Continental rises	6%	31%	8 km (5 mi)
Deep-ocean floor	78%	13%	0.6 km (0.4 mi)

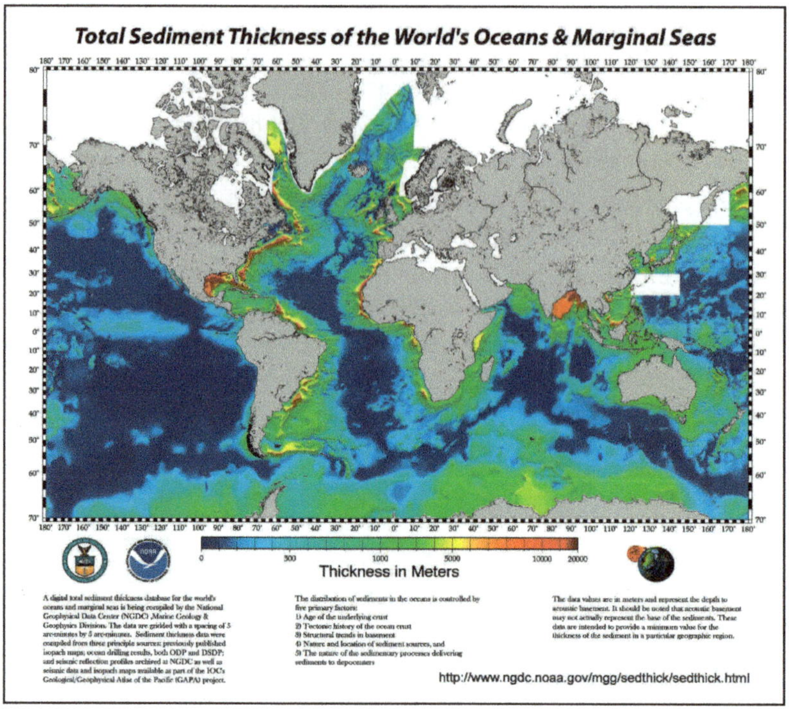

Total sediment thickness of the ocean floor, with the thinnest deposits in dark blue and the thickest in red. Note the abundant deposits along the east and gulf coasts of North America, in the South China Sea, and in the Bay of Bengal, east of India

Classification of Marine Sediments by Source of Particles

Sediment Type	Source	Examples	Distribution	Percent of All Ocean Floor Area Covered
Terrigenous	Erosion of land, volcanic eruptions, blown dust	Quartz sand, clays, estuarine mud	Dominant on continental margins, abyssal plains, polar ocean floors	~45%
Biogenous	Organic; accumulation of hard parts of some marine organisms	Calcareous and siliceous oozes	Dominant on deep-ocean floor (siliceous ooze below about 5 km)	~55%
Hydrogenous (authigenic)	Precipitation of dissolved mineral from water, often by bacteria	Manganese nodules, phosphorite deposits	Present with other, more dominant sediments	1%
Cosmogenous	Dust from space, meteorite debris	Tektite spheres, glassy nodules	Mixed in very small proportion with more dominant sediments	1%

PELAGIC RED CLAY

Pelagic clay accumulates in the deepest and most remote areas of the ocean. It covers 38% of the ocean floor and accumulates more slowly than any other sediment type, at only 0.1–0.5 cm/1000 yr. Containing less than 30% biogenic material, it consists of sediment that remains after the dissolution of both calcareous and siliceous biogenic particles while they settled through the water column. These sediments consist of eolian quartz, clay minerals, volcanic ash, subordinate residue of siliceous microfossils, and authigenic minerals such as zeolites, limonite and manganese oxides. The bulk of red clay consists of eolian dust. Accessory constituents found in red clay include meteorite dust, fish bones and teeth, whale ear bones, and manganese micro-nodules.

These pelagic sediments are typically bright red to chocolate brown in color. The color results from coatings of iron oxide and manganese oxide on the sediment particles. In the absence of organic carbon, iron and manganese remain in their oxidized states and these clays remain brown after burial. When more deeply buried, brown clay may change into red clay due to the conversion of iron hydroxides to hematite.

These sediments accumulate on the ocean floor within areas characterized by little planktonic production. The clays which comprise them were transported into the deep ocean in suspension, either in the air over the oceans or in surface waters. Both wind and ocean currents transported these sediments in suspension thousands of kilometers from their terrestrial source. As they were transported, the finer clays may have stayed in suspension for a hundred years or more within the water column before they settled to the ocean bottom. The settling of this clay-size sediment occurred primarily by the formation of clay aggregates by flocculation and by their incorporation into fecal pellets by pelagic organisms.

REFERENCES

1. Rothwell, R.G., (2005) Deep Ocean Pelagic Oozes, Vol. 5. of Selley, Richard C., L. Robin McCocks, and Ian R. Plimer, Encyclopedia of Geology, Oxford: Elsevier Limited. ISBN 0-12-636380-3
2. HüNeke, H., and T. Mulder (2011) Deep-Sea Sediments. Developments in Sedimentology, vol. 63. Elsiever, New York. 849 pp. ISBN 978-0-444-53000-4
3. Neuendorf, K.K.E., J.P. Mehl, Jr., and J.A. Jackson, J.A., eds. (2005) Glossary of Geology (5th ed.). Alexandria, Virginia, American Geological Institute. 779 pp. ISBN 0-922152-76-4

CHAPTER 6

Physical Oceanography

Chapter Contents

- **Introduction**
- **Circulation**
- **Ocean Current**

The Physical Oceanography group at Horn Point Laboratory spans a diverse range of interests and research. Generally physical oceanographers are concerned with the motion of the ocean. This includes waves, currents, movement and erosion of sediments, pollutants and biology, interactions of the ocean with the atmosphere and the land surface, and the interactions of the ocean with climate variability.

INTRODUCTION

Physical oceanography is the study of physical conditions and physical processes within the ocean, especially the motions and physical properties of ocean waters.

Physical oceanography is one of several sub-domains into which oceanography is divided. Others include biological, chemical and geological oceanography.

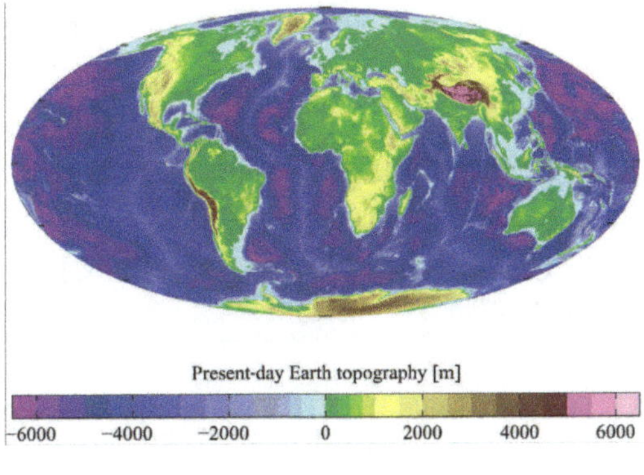

World ocean bathymetry.

Physical Setting

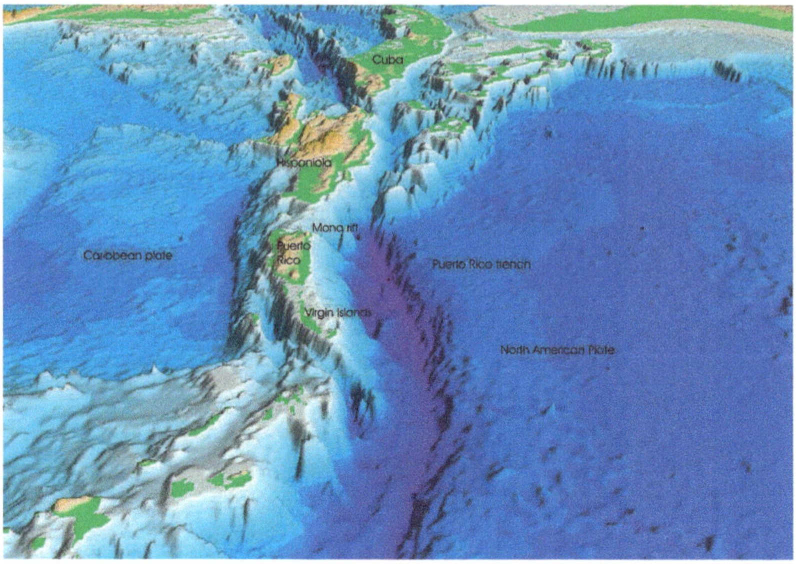

Perspective view of the sea floor of the Atlantic Ocean and the Caribbean Sea. The purple sea floor at the center of the view is the Puerto Rico Trench.

The pioneering oceanographer Matthew Maury said in 1855 *"Our planet is invested with two great oceans; one visible, the other invisible; one underfoot, the other overhead; one entirely envelopes it, the other covers about two thirds of its surface."* The fundamental role of the oceans in shaping Earth is acknowledged by ecologists, geologists, meteorologists, climatologists, geographers and others interested in the physical world. An Earth without oceans would truly be unrecognizable.

Roughly 97% of the planet›s water is in its oceans, and the oceans are the source of the vast majority of water vapor that condenses in the atmosphere and falls as rain or snow on the continents. The tremendous heat capacity of the oceans moderates the planet›s climate, and its absorption of various gases affects the composition of the atmosphere. The ocean›s influence extends even to the composition of volcanic rocks through seafloor metamorphism, as well as to that of volcanic gases and magmas created at subduction zones.

The oceans are far deeper than the continents are tall; examination of the Earth›s hypsographic curve shows that the average elevation of Earth›s landmasses is only 840 metres (2,760 ft), while the ocean›s average depth is 3,800 metres (12,500 ft). Though this apparent discrepancy is great, for both land and sea, the respective extremes such as mountains and trenches are rare.

Area, volume plus mean and maximum depths of oceans (excluding adjacent seas)

Body	**Area** ($10^6 km^2$)	**Volume** ($10^6 km^3$)	**Mean depth** (m)	**Maximum** (m)

Pacific Ocean	165.2	707.6	4282	-11033
Atlantic Ocean	82.4	323.6	3926	-8605
Indian Ocean	73.4	291.0	3963	-8047
Southern Ocean	20.3			-7235
Arctic Ocean	14.1		1038	
Caribbean Sea	2.8			-7686

Temperature, Salinity and Density

Because the vast majority of the world ocean's volume is deep water, the mean temperature of seawater is low; roughly 75% of the ocean's volume has a temperature from 0° – 5 °C (Pinet 1996). The same percentage falls in a salinity range between 34–35 ppt (3.4–3.5%) (Pinet 1996). There is still quite a bit of variation, however. Surface temperatures can range from below freezing near the poles to 35 °C in restricted tropical seas, while salinity can vary from 10 to 41 ppt (1.0–4.1%).

The vertical structure of the temperature can be divided into three basic layers, a surface mixed layer, where gradients are low, a thermocline where gradients are high, and a poorly stratified abyss.

In terms of temperature, the ocean's layers are highly latitude-dependent; the thermocline is pronounced in the tropics, but nonexistent in polar waters (Marshak 2001). The halocline usually lies near the surface, where evaporation raises salinity in the tropics, or meltwater dilutes it in polar regions. These variations of salinity and temperature with depth change the density of the seawater, creating the pycnocline.

CIRCULATION

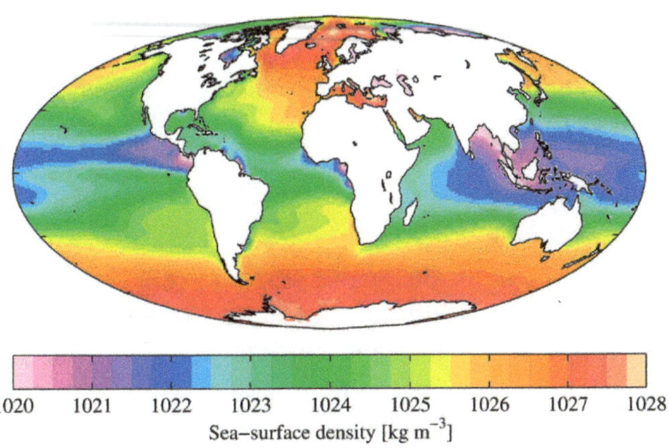

Density-driven thermohaline circulation

Energy for the ocean circulation (and for the atmospheric circulation) comes from solar radiation and gravitational energy from the sun and moon. The amount of sunlight absorbed at the surface varies strongly with latitude, being greater at the equator than at the poles, and this engenders fluid motion in both the atmosphere and ocean that acts to redistribute heat from the equator towards the poles, thereby reducing the temperature gradients that would exist in the absence of fluid motion. Perhaps three quarters of this heat is carried in the atmosphere; the rest is carried in the ocean.

The atmosphere is heated from below, which leads to convection, the largest expression of which is the Hadley circulation. By contrast the ocean is heated from above, which tends to suppress convection. Instead ocean deep water is formed in polar regions where cold salty waters sink in fairly restricted areas. This is the beginning of the thermohaline circulation.

Oceanic currents are largely driven by the surface wind stress; hence the large-scale atmospheric circulation is important to understanding the ocean circulation. The Hadley circulation leads to Easterly winds in the tropics and Westerlies in mid-latitudes. This leads to slow equatorward flow throughout most of a subtropical ocean basin (the Sverdrup balance). The return flow occurs in an intense, narrow, poleward western boundary current. Like the atmosphere, the ocean is far wider than it is deep, and hence horizontal motion is in general much faster than vertical motion. In the southern hemisphere there is a continuous belt of ocean, and hence the mid-latitude westerlies force the strong Antarctic Circumpolar Current. In the northern hemisphere the land masses prevent this and the ocean circulation is broken into smaller gyres in the Atlantic and Pacific basins.

Coriolis Effect

The Coriolis effect results in a deflection of fluid flows (to the right in the Northern Hemisphere and left in the Southern Hemisphere). This has profound effects on the flow of the oceans. In particular it means the flow goes *around* high and low pressure systems, permitting them to persist for long periods of time. As a result, tiny variations in pressure can produce measurable currents. A slope of one part in one million in sea surface height, for example, will result in a current of 10 cm/s at mid-latitudes. The fact that the Coriolis effect is largest at the poles and weak at the equator results in sharp, relatively steady western boundary currents which are absent on eastern boundaries. Also see secondary circulation effects.

Ekman Transport

Ekman transport results in the net transport of surface water 90 degrees to the right of the wind in the Northern Hemisphere, and 90 degrees to the left of the wind in the Southern Hemisphere. As the wind blows across the surface of the ocean, it "grabs" onto a thin layer of the surface water. In turn, that thin sheet of water transfers motion energy to the thin layer of water under it, and so on. However, because of the Coriolis Effect, the direction of travel of the layers of water slowly move farther and farther to the right as they get deeper in the Northern Hemisphere, and to the left in the Southern Hemisphere. In most cases, the very bottom layer of water affected by the wind is at a depth of 100 m – 150 m and is traveling about 180 degrees, completely opposite of the direction that the wind is blowing. Overall, the net transport of water would be 90 degrees from the original direction of the wind.

Langmuir Circulation

Langmuir circulation results in the occurrence of thin, visible stripes, called windrows on the surface of the ocean parallel to the direction that the wind is blowing. If the wind is blowing with more than 3 m s^{-1}, it can create parallel windrows alternating upwelling and downwelling about 5–300 m apart. These windrows are created by adjacent ovular water cells (extending to about 6 m (20 ft) deep) alternating rotating clockwise and counterclockwise. In the convergence zones debris, foam and seaweed accumulates, while at the divergence zones plankton are caught and carried to the surface. If there are many plankton in the divergence zone fish are often attracted to feed on them.

Ocean–Atmosphere Interface

Hurricane Isabel east of the Bahamas on 15 September 2003

At the ocean-atmosphere interface, the ocean and atmosphere exchange fluxes of heat, moisture and momentum.

Heat

The important heat terms at the surface are the sensible heat flux, the latent heat flux, the incoming solar radiation and the balance of long-wave (infrared) radiation. In general, the tropical oceans will tend to show a net gain of heat, and the polar oceans a net loss, the result of a net transfer of energy polewards in the oceans.

The oceans' large heat capacity moderates the climate of areas adjacent to the oceans, leading to a maritime climate at such locations. This can be a result of heat storage in summer and release in winter; or of transport of heat from warmer locations: a particularly notable example of this is Western Europe, which is heated at least in part by the north atlantic drift.

Momentum

Surface winds tend to be of order meters per second; ocean currents of order centimeters per second. Hence from the point of view of the atmosphere, the ocean can be considered effectively stationary; from the point of view of the ocean, the atmosphere imposes a significant wind stress on its surface, and this forces large-scale currents in the ocean.

Through the wind stress, the wind generates ocean surface waves; the longer waves have a phase velocity tending towards the wind speed. Momentum of the surface winds is transferred into the energy flux by the ocean surface waves. The increased roughness of the ocean surface, by the presence of the waves, changes the wind near the surface.

Moisture

The ocean can gain moisture from rainfall, or lose it through evaporation. Evaporative loss leaves the ocean saltier; the Mediterranean and Persian Gulf for example have strong evaporative loss; the resulting plume of dense salty water may be traced through the Straits of Gibraltar into the Atlantic Ocean. At one time, it was believed that evaporation/precipitation was a major driver of ocean currents; it is now known to be only a very minor factor.

Planetary Waves

Kelvin Waves

A Kelvin wave is any progressive wave that is channeled between two boundaries or opposing forces (usually between the Coriolis force and a coastline or the equator). There are two types, coastal and equatorial. Kelvin waves are gravity driven and non-dispersive. This means that Kelvin waves can retain their shape and direction over long periods of time. They are usually created by a sudden shift in the wind, such as the change of the trade winds at the beginning of the El Niño-Southern Oscillation.

Coastal Kelvin waves follow shorelines and will always propagate in a counterclockwise direction in the Northern hemisphere (with the shoreline to the right of the direction of travel) and clockwise in the Southern hemisphere. Equatorial Kelvin waves propagate to the east in the Northern and Southern hemispheres, using the equator as a guide.

Kelvin waves are known to have very high speeds, typically around 2–3 meters per second. They have wavelengths of thousands of kilometers and amplitudes in the tens of meters.

Rossby Waves

Rossby waves, or planetary waves are huge, slow waves generated in the troposphere by temperature differences between the ocean and the continents. Their major restoring force is the change in Coriolis force with latitude. Their wave amplitudes are usually in the tens of meters and very large wavelengths. They are usually found at low or mid latitudes.

There are two types of Rossby waves, barotropic and baroclinic. Barotropic Rossby waves have the highest speeds and do not vary vertically. Baroclinic Rossby waves are much slower.

The special identifying feature of Rossby waves is that the phase velocity of each individual wave always has a westward component, but the group velocity can be in any direction. Usually the shorter Rossby waves have an eastward group velocity and the longer ones have a westward group velocity.

Climate Variability

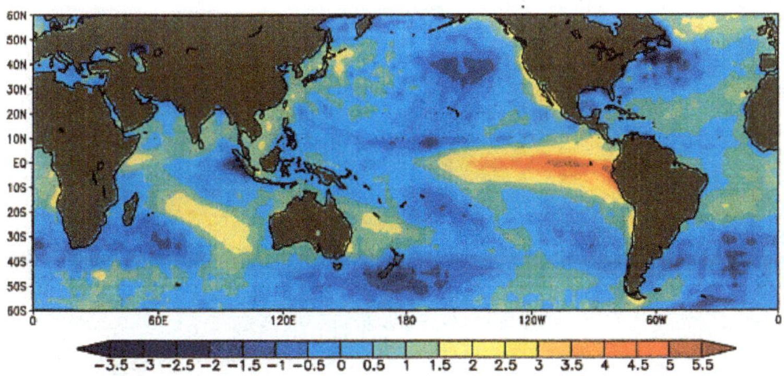

December 1997 chart of ocean surface temperature anomaly °C during the last strong El Niño

The interaction of ocean circulation, which serves as a type of heat pump, and biological effects such as the concentration of carbon dioxide can result in global climate changes on a time scale of decades. Known climate oscillations resulting from these interactions, include the Pacific decadal oscillation, North Atlantic oscillation, and Arctic oscillation. The oceanic process of thermohaline circulation is a significant component of heat redistribution across the globe, and changes in this circulation can have major impacts upon the climate.

La Niña–El Niño

Antarctic Circumpolar Wave

This is a coupled ocean/atmosphere wave that circles the Southern Ocean about every eight years. Since it is a wave-2 phenomenon (there are two peaks and two troughs in a latitude circle) at each fixed point in space a signal with a period of four years is seen. The wave moves eastward in the direction of the Antarctic Circumpolar Current.

Ocean Currents

Among the most important ocean currents are the:

- Antarctic Circumpolar Current
- Deep ocean (density-driven)
- Western boundary currents
- Gulf Stream
- Kuroshio Current
- Labrador Current
- Oyashio Current
- Agulhas Current
- Brazil Current
- East Australia Current
- Eastern Boundary currents
- California Current
- Canary Current
- Peru Current
- Benguela Current

Antarctic Circumpolar

The ocean body surrounding the Antarctic is currently the only continuous body of water where there is a wide latitude band of open water. It interconnects the Atlantic, Pacific and Indian oceans, and provide an uninterrupted stretch for the prevailing westerly winds to significantly increase wave amplitudes. It is generally accepted that these prevailing winds are primarily responsible for the circumpolar current transport. This current is now thought to vary with time, possibly in an oscillatory manner.

Deep Ocean

In the Norwegian Sea evaporative cooling is predominant, and the sinking water mass, the North Atlantic Deep Water (NADW), fills the basin and spills southwards through crevasses in the submarine sills that connect Greenland, Iceland and Britain. It then flows along the western boundary of the Atlantic with some part of the flow moving eastward along the equator and then poleward into the ocean basins. The NADW is entrained into the Circumpolar Current, and can be traced into the Indian and Pacific basins. Flow from the Arctic Ocean Basin into the Pacific, however, is blocked by the narrow shallows of the Bering Strait.

Also see **marine geology** about that explores the geology of the ocean floor including plate tectonics that create deep ocean trenches.

Western Boundary

An idealised subtropical ocean basin forced by winds circling around a high pressure (anticyclonic) systems such as the Azores-Bermuda high develops a gyre circulation with slow steady flows towards the equator in the interior. As discussed by Henry Stommel, these flows are balanced in the region of the western boundary, where a thin fast polewards flow called a western boundary current develops. Flow in the real ocean is more complex, but the Gulf stream, Agulhas and Kuroshio are examples of such currents. They are narrow (approximately 100 km across) and fast (approximately 1.5 m/s).

Equatorwards western boundary currents occur in tropical and polar locations, e.g. the East Greenland and Labrador currents, in the Atlantic and the Oyashio. They are forced by winds circulation around low pressure (cyclonic).

Gulf stream

The Gulf Stream, together with its northern extension, North Atlantic Current, is a powerful, warm, and swift Atlantic Ocean current that originates in the Gulf of Mexico, exits through the Strait of Florida, and follows the eastern coastlines of the United States and Newfoundland to the northeast before crossing the Atlantic Ocean.

Kuroshio

The Kuroshio Current is an ocean current found in the western Pacific Ocean off the east coast of Taiwan and flowing northeastward past Japan, where it merges with the easterly drift of the North Pacific Current. It is analogous to the Gulf Stream in the Atlantic Ocean, transporting warm, tropical water northward towards the polar region.

Heat Flux

Heat Storage

Ocean heat flux is a turbulent and complex system which utilizes atmospheric measurement techniques such as eddy covariance to measure the rate of heat transfer expressed in the unit of joules or watts per second. Heat flux is the difference in temperature between two points through which the heat passes. Most of the Earth's heat storage is within its seas with smaller fractions of the heat transfer in processes such as evaporation, radiation, diffusion, or absorption into the sea floor. The majority of the ocean heat flux is through advection or the movement of the ocean's currents. For example, the majority of the warm water movement in the south Atlantic is thought to have originated in the Indian Ocean. Another example of advection is the nonequatorial Pacific heating which results from subsurface processes related to atmospheric anticlines. Recent warming observations of Antarctic Bottom Water in the Southern Ocean is of concern to ocean scientists because bottom water changes will effect currents, nutrients, and biota elsewhere. The international awareness of global warming has focused scientific research on this topic since the 1988 creation of the Intergovernmental Panel on Climate Change. Improved ocean observation, instrumentation, theory, and funding has increased scientific reporting on regional and global issues related to heat.

Rapid Variations

Tides

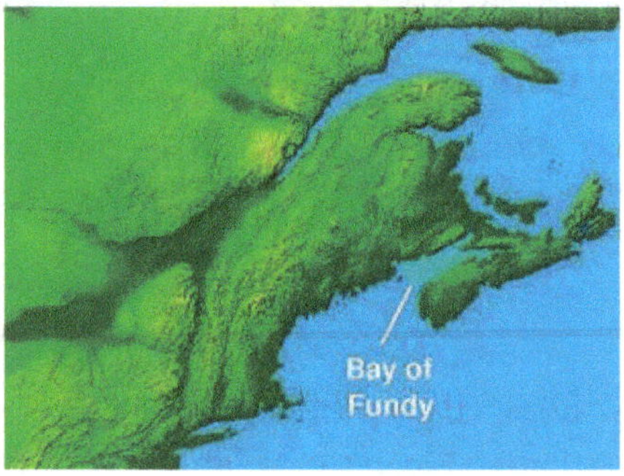

The **Bay of Fundy** is a bay located on the Atlantic coast of North America, on the northeast end of the Gulf of Maine between the provinces of New Brunswick and Nova Scotia.

The rise and fall of the oceans due to tidal effects is a key influence upon the coastal areas. Ocean tides on the planet Earth are created by the gravitational effects of the Sun and Moon. The tides produced by these two bodies are roughly comparable in magnitude, but the orbital motion of the Moon results in tidal patterns that vary over the course of a month.

The ebb and flow of the tides produce a cyclical current along the coast, and the strength of this current can be quite dramatic along narrow estuaries. Incoming tides can also produce a tidal bore along a river or narrow bay as the water flow against the current results in a wave on the surface.

Tide and Current (Wyban 1992) clearly illustrates the impact of these natural cycles on the lifestyle and livelihood of Native Hawaiians tending coastal fishponds. *Aia ke ola ka hana* meaning . . . *Life is in labor*.

Tidal resonance occurs in the Bay of Fundy since the time it takes for a large wave to travel from the mouth of the bay to the opposite end, then reflect and travel back to the mouth of the bay coincides with the tidal rhythm producing the world's highest tides.

As the surface tide oscillates over topography, such as submerged seamounts or ridges, it generates internal waves at the tidal frequency, which are known as internal tides.

Tsunamis

A series of surface waves can be generated due to large-scale displacement of the ocean water. These can be caused by sub-marine landslides, seafloor deformations due to earthquakes, or the impact of a large meteorite.

The waves can travel with a velocity of up to several hundred km/hour across the ocean surface, but in mid-ocean they are barely detectable with wavelengths spanning hundreds of kilometers.

Tsunamis, originally called tidal waves, were renamed because they are not related to the tides. They are regarded as shallow-water waves, or waves in water with a depth less than 1/20 their wavelength. Tsunamis have very large periods, high speeds, and great wave heights.

The primary impact of these waves is along the coastal shoreline, as large amounts of ocean water are cyclically propelled inland and then drawn out to sea. This can result in significant modifications to the coastline regions where the waves strike with sufficient energy.

The tsunami that occurred in Lituya Bay, Alaska on July 9, 1958 was 520 m (1,710 ft) high and is the biggest tsunami ever measured, almost 90 m (300 ft) taller than the Sears Tower in Chicago and about 110 m (360 ft) taller than the former World Trade Center in New York.

Surface Waves

The wind generates ocean surface waves, which have a large impact on offshore structures, ships, coastal erosion and sedimentation, as well as harbours. After their generation by the wind, ocean surface waves can travel (as swell) over long distances.

OCEAN CURRENT

An **ocean current** is a continuous, directed movement of seawater generated by forces acting upon this mean flow, such as breaking waves, wind, the Coriolis effect, cabbeling, temperature and salinity differences, while tides are caused by the gravitational pull of the Sun and Moon. Depth contours, shoreline configurations, and interactions with other currents influence a current›s direction and strength.

Ocean currents flow for great distances, and together, create the global conveyor belt which plays a dominant role in determining the climate of many of the Earth's regions. More specifically, ocean currents influence the temperature of the regions through which they travel. For example, warm currents traveling along more temperate coasts increase the temperature of the area by warming the sea breezes that blow over them. Perhaps the most striking example is the Gulf Stream, which makes northwest Europe much more temperate than any other region at the same latitude. Another example is Lima, Peru where the climate is cooler (sub-tropical) than the tropical latitudes in which the area is located, due to the effect of the Humboldt Current.

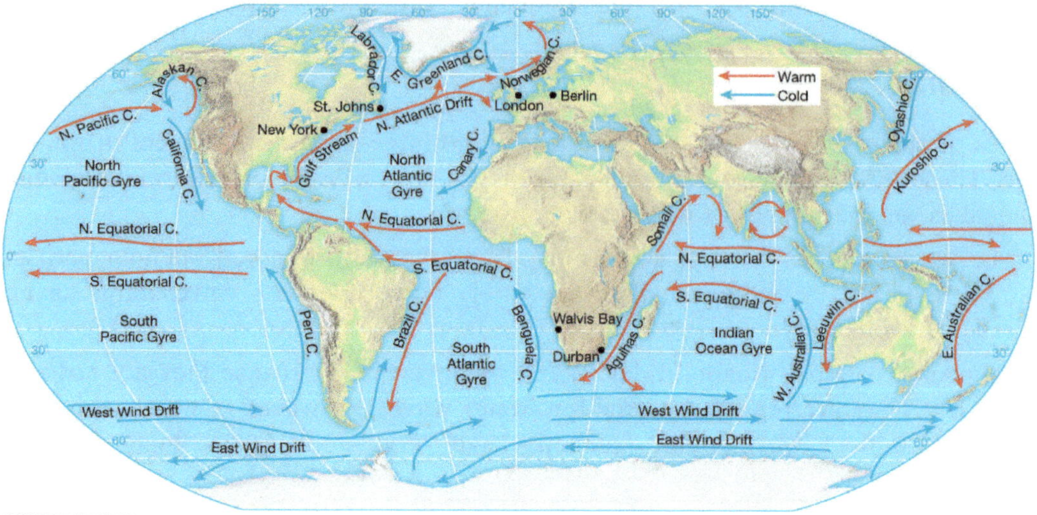

© 2013 Pearson Education, Inc.

The ocean currents.

Function

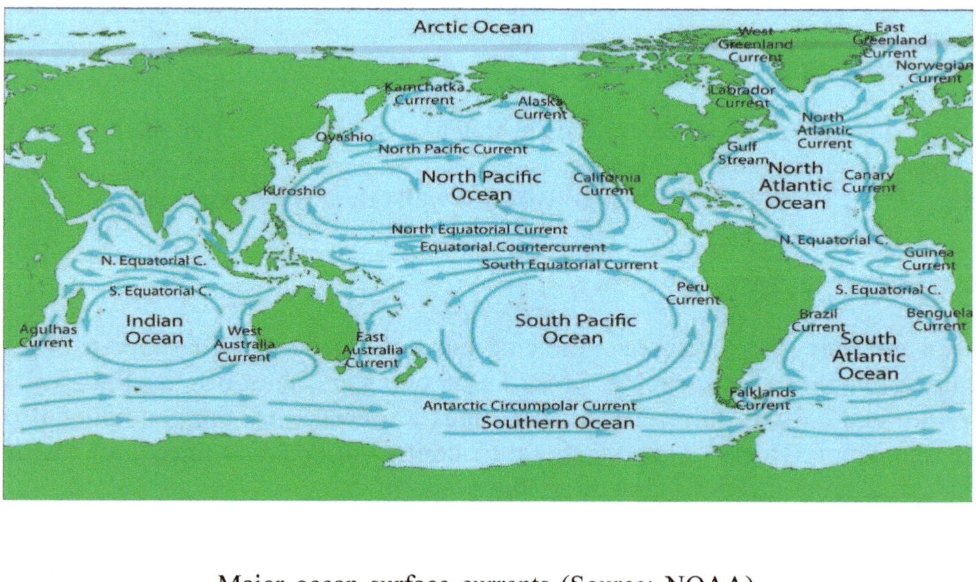

Major ocean surface currents (Source: NOAA)

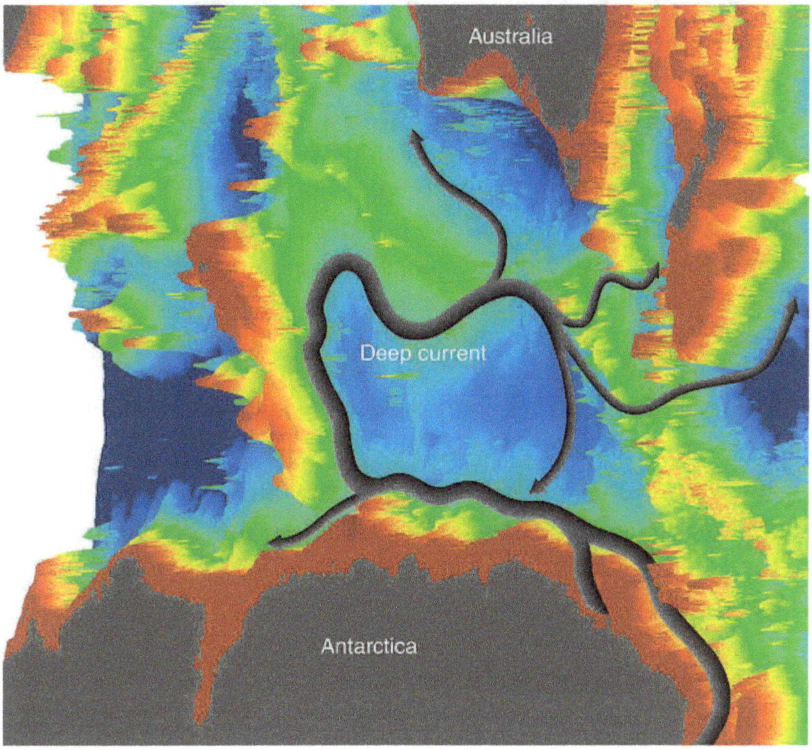

The bathymetry of the Kerguelen Plateau in the Southern Ocean governs the course of the new current part of the global network of ocean currents.

Surface oceanic currents are sometimes wind driven and develop their typical clockwise spirals in the northern hemisphere and counter-clockwise rotation in the southern hemisphere due to imposed wind stresses. In these wind driven currents, the Ekman spiral effect results in the currents flowing at an angle to the driving winds. In addition, the areas of surface ocean currents move somewhat with the seasons; this is most notable in equatorial currents.

Deep ocean basins generally have a non-symmetric surface current, in that the eastern equatorward-flowing branch is broad and diffuse whereas the western poleward flowing branch is very narrow. These western boundary currents (of which the Gulf Stream is an example) are a consequence of the rotation of the Earth.

Deep ocean currents are driven by density and temperature gradients. Thermohaline circulation is also known as the ocean's conveyor belt (which refers to deep ocean density driven ocean basin currents). These currents, called submarine rivers, flow under the surface of the ocean and are hidden from immediate detection. Where significant vertical movement of ocean currents is observed, this is known as upwelling and downwelling. Deep ocean currents are currently being researched using a fleet of underwater robots called Argo.

The South Equatorial Currents of the Atlantic and Pacific straddle the equator. Though the Coriolis effect is weak near the equator (and absent at the equator), water moving in the currents on either side of the equator is deflected slightly poleward and replaced by deeper water. Thus, equatorial upwelling occurs in these westward flowing equatorial surface currents. Upwelling is an important process because this water from within and below the pycnocline is often rich in nutrients and greatly benefits the growth of marine organisms. By contrast, generally poor conditions for growth prevail in most of the open tropical ocean because strong layering isolates deep, nutrient rich water from the sunlit ocean surface.

Surface currents make up only 8% of all water in the ocean, are generally restricted to the upper 400 m (1,300 ft) of ocean water, and are separated from lower regions by varying temperatures and salinity which affect the density of the water, which in turn, defines each oceanic region. Because the movement of deep water in ocean basins is caused by density driven forces and gravity, deep waters sink into deep ocean basins at high latitudes where the temperatures are cold enough to cause the density to increase.

Ocean currents are measured in sverdrup (sv), where 1 sv is equivalent to a volume flow rate of 1,000,000 m^3(35,000,000 cu ft) per second.

Surface currents are found on the surface of an ocean, and are driven by large scale wind currents. They are directly affected by the wind—the Coriolis effect plays a role in their behaviors.

Thermohaline Circulation

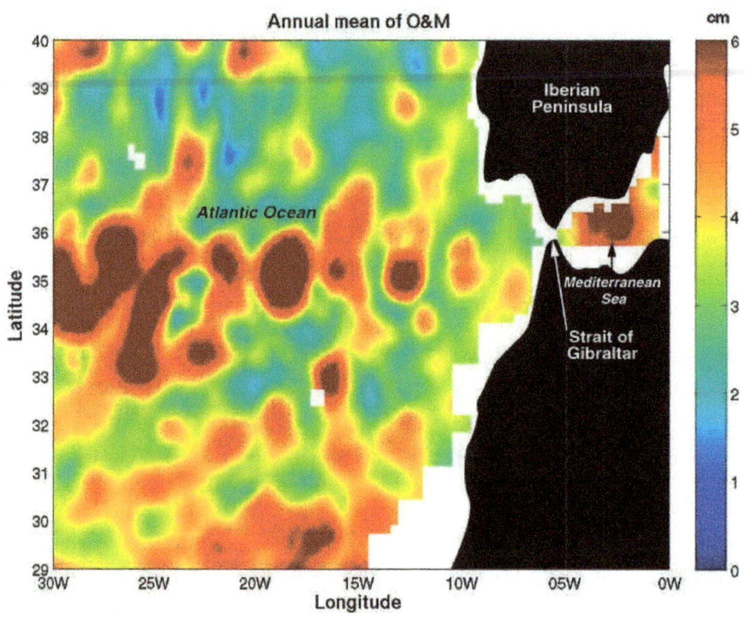

Coupling data collected by NASA/JPL by several different satellite-borne sensors, researchers have been able to "break through" the ocean's surface to detect "Meddies" -- super-salty warm-water eddies that originate in the Mediterranean Sea and then sink more than a half-mile underwater in the Atlantic Ocean. The Meddies are shown in red in this scientific figure. Horizontal and vertical currents also exist below the pycnocline in the ocean's deeper waters. The movement of water due to differences in density as a function of water temperature and salinity is called thermohaline circulation. Ripple marks in sediments, scour lines, and the erosion of rocky outcrops on deep-ocean floors are evidence that relatively strong, localized bottom currents exist. Some of these currents may move as rapidly as 60 centimeters (24 inches) per second.

These currents are strongly influenced by bottom topography, since dense, bottom water must forcefully flow over and around depressions and projections in the seafloor. Thus, they are sometimes called contour currents. Bottom currents generally move equator-ward at or near the western boundaries of ocean basins (below the western boundary surface currents). The deep-water masses are not capable of moving water at speeds comparable to that of wind-driven surface currents. Water in some of these currents may move only 1 to 2 meters per day; however, even at that slow speed, the Coriolis effect modifies their pattern of flow.

Downwelling of Deep Water in Polar Regions

A recording current meter

Antarctic Bottom Water is the most distinctive of the deep-water masses. It is characterized by a salinity of 34.65‰, a temperature of -0.5 °C (30 °F), and a density of 1.0279 grams per cubic centimeter. This water is noted for its extreme density (the densest in the world ocean), for the great amount of it produced near Antarctic coasts, and for its ability to migrate north along the seafloor. Most Antarctic Bottom Water forms near the Antarctic coast south of South America during winter. Salt is concentrated in pockets between crystals of pure water and then squeezed out of the freezing mass to form a frigid brine. Between 20 million and 50 million cubic meters of this brine form every second. The water's great density causes it to sink toward the continental shelf, where it mixes with nearly equal parts of water from the southern Antarctic Circumpolar Current. The mixture settles along the edge of Antarctica's continental shelf, descends along the slope, and spreads along the deep-sea bed, creeping north in slow sheets. Antarctic Bottom Water flows many times as slowly as the water in surface currents: in the Pacific it may take a thousand years to reach the equator. Antarctic Bottom Water also flows into the Atlantic Ocean basin, where it flows north at a faster rate than in the Pacific. Antarctic Bottom Water has been identified as high as 40° N on the Atlantic floor.

A small amount of dense bottom water also forms in the northern polar ocean. Although, the topography of the Arctic Ocean basin prevents most of the bottom water from escaping, with the exception of deep channels formed in the submarine ridges between Scotland, Iceland, and Greenland. These channels allow the cold, dense water formed in the Arctic to flow into the North Atlantic to form North Atlantic Deep Water. North Atlantic Deep Water forms when the relatively warm and salty North Atlantic Ocean cools as cold winds from northern Canada sweep over it. Exposed to the chilled air, water at the latitude of Iceland releases heat, cools from 10 °C to 2 °C, and sinks. Gulf Stream water that sinks in the north is replaced by warm water flowing clockwise along the U.S. east coast in the North Atlantic gyre.

Importance

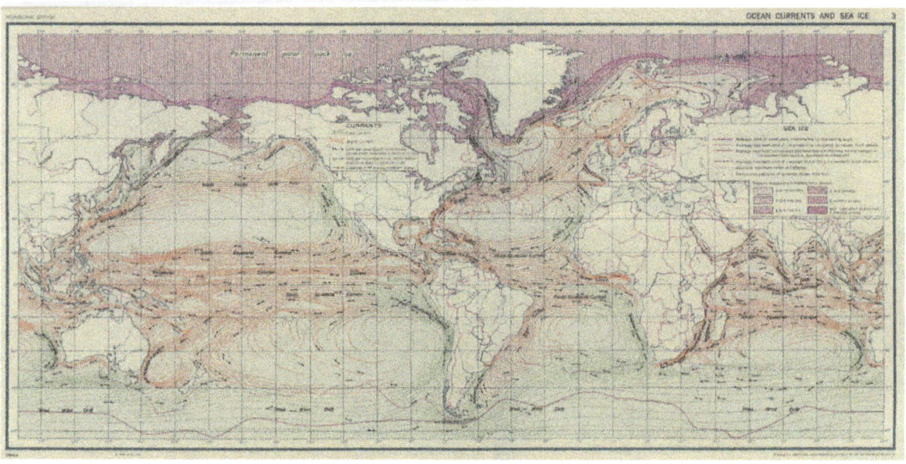

A 1943 map of the world's ocean currents.

Knowledge of surface ocean currents is essential in reducing costs of shipping, since traveling with them reduces fuel costs. In the wind powered sailing-ship era, knowledge of wind patterns and ocean currents was even more essential. A good example of this is the Agulhas Current, which long prevented Portuguese sailors from reaching India. In recent times, around-the-world sailing competitors make good use of surface currents to build and maintain speed. Ocean currents are also very important in the dispersal of many life forms. An example is the life-cycle of the European Eel.

Ocean currents are important in the study of marine debris, and vice versa. These currents also affect temperatures throughout the world. For example, the ocean current that brings warm water up the north Atlantic to northwest Europe also cumulatively and slowly blocks ice from forming along the seashores, which would also block ships from entering and exiting inland waterways and seaports, hence ocean currents play a decisive role in influencing the climates of regions through which they flow. Cold ocean water currents flowing from polar and sub-polar regions bring in a lot of plankton that are crucial to the continued survival of several key sea creature species in marine ecosystems. Since plankton are the food of fish, abundant fish populations often live where these currents prevail.

Ocean currents can also be used for marine power generation, with areas off of Japan, Florida and Hawaii being considered for test projects.

OSCAR: Near-Realtime Global Ocean Surface Current Data Set

The OSCAR Near-realtime global ocean surface currents website describes the project and links to data validation and data downloads.

OSCAR data is used extensively in climate studies. maps and descriptions or annotations of climatic anomalies have been published in the monthly Climate Diagnostic Bulletin since 2001 and are routinely used to monitor ENSO and to test weather prediction models. OSCAR currents are routinely used to evaluate the surface currents in Global Circulation Models (GCMs), for example in NCEP Global Ocean Data Assimilation System (GODAS) and European Centre for Medium-Range Weather Forecasts (ECMWF).

REFERENCES

1. Physical Oceanography Oregon State University.
2. Pinet, Paul R. (1996). Invitation to Oceanography (3rd ed.). St. Paul, MN: West Publishing Co. ISBN 0-7637-2136-0.
3. Hamblin, W. Kenneth; Christiansen, Eric H. (1998). Earth's Dynamic Systems (8th ed.). Upper Saddle River: Prentice-Hall. ISBN 0-13-018371-7.
4. Marshak, Stephen (2001). Earth: Portrait of a Planet. New York: W.W. Norton & Company. ISBN 0-393-97423-5.
5. Munk, W. and Wunsch, C., 1998: Abyssal recipes II: energetics of tidal and wind mixing. Deep-Sea Research Part I, 45, pp. 1977--2010.
6. Talley, Lynne D. "Reading-Advection, transports, budgets". SIO 210: Introduction to Physical Oceanography. (Fall 2013). San Diego: Scripps Institute of Oceanography. University of California San Diego. Retrieved August 30, 2014. http://www-pord.ucsd.edu/~ltalley/sio210/transports/
7. Macdonald, Alison M.(1995). Oceanic fluxes of mass, heat, and freshwater :

a global estimate and perspective. Falmouth, Mass.: Massachusetts Institute of Technology and the Woods Hole Oceanographic Institution. p. 12. WHOI Theses. Retrieved August 30, 2014. http://hdl.handle.net/1912/5620

8. Su, Jingzhi; Li, Tim; et al. (Jun 2014). "The Initiation and Developing Mechanisms of Central Pacific El Niños." Journal of Climate 27(12):4473-4485. DOI: 10.1175/JCLI-D-13-00640.1
9. Goldman, Jana. (March 20, 2012). Amount of coldest Antarctic water near ocean floor decreasing for decades. NOAA. Retrieved 30 August 2014. http://www.noaanews.noaa.gov/stories2012/20120320_antarcticbottomwater.html
10. WorldCat. "MyWorldCat list-OceanHeat". Retrieved Aug 30, 2014. https://www.worldcat.org/profiles/greycloud/lists/3461529
11. Stocker, Thomas F. (2013). Technical Summary In: Climate Change 2013: The Physical Science Basis. Contribution of Working Group I to the Fifth Assessment Report of the Intergovernmental Panel on Climate Change. Cambridge: Cambridge University Press. p. 90.
12. "Tsuanmi Threats". Retrieved 2008-06-28.

CHAPTER 7

Wind Wave

Chapter Contents

The wind, wave and weather forecast for Clinton/Truman Lake in Missouri, United States. Windfinder specializes in wind, waves, tides and weather reports & forecasts for wind related sports like kitesurfing, windsurfing, surfing, sailing or paragliding.

INTRODUCTION

In fluid dynamics, wind waves, or wind-generated waves, are surface waves that occur on the free surface of oceans, seas, lakes, rivers, and canals or even on small puddles and ponds. They result from the wind blowing over an area of fluid surface. Waves in the oceans can travel thousands of miles before reaching land. Wind waves range in size from small ripples, to waves over 100 ft (30 m) high.

When directly generated and affected by local winds, a wind wave system is called a wind sea. After the wind ceases to blow, wind waves are called swells. More generally, a swell consists of wind-generated waves that are not significantly affected by the local wind at that time. They have been generated elsewhere or some time ago. Wind waves in the ocean are called ocean surface waves.

Wind waves have a certain amount of randomness: subsequent waves differ in height, duration, and shape with limited predictability. They can be described as a stochastic process, in combination with the physics governing their generation, growth, propagation and decay—as well as governing the interdependence between flow quantities such as: the water surface movements, flow velocities and water pressure. The key statistics of wind waves (both seas and swells) in evolving sea states can be predicted with wind wave models.

Although waves are usually considered in the water seas of Earth, the hydrocarbon seas of Titan may also have wind-driven waves.

Large wave

Ocean waves

WAVE FORMATION

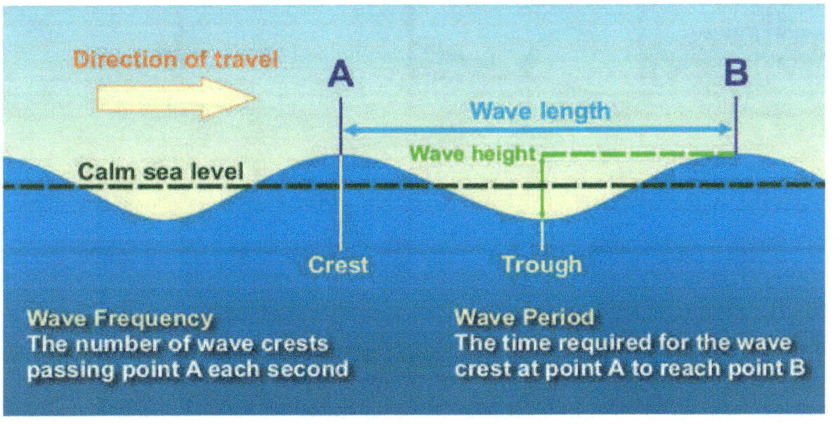

Aspects of a water wave

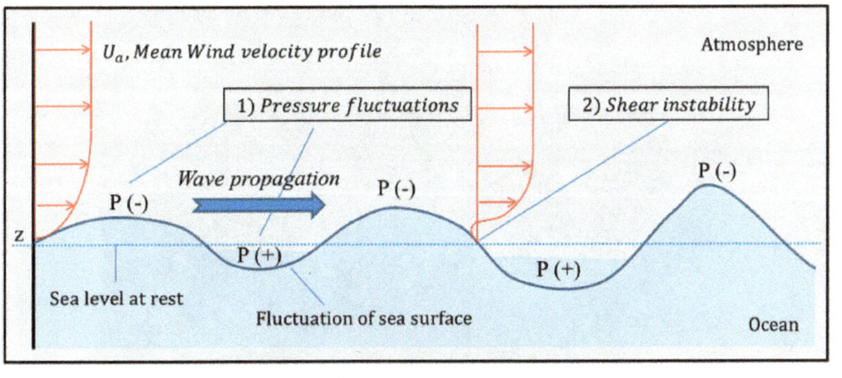

Wave formation

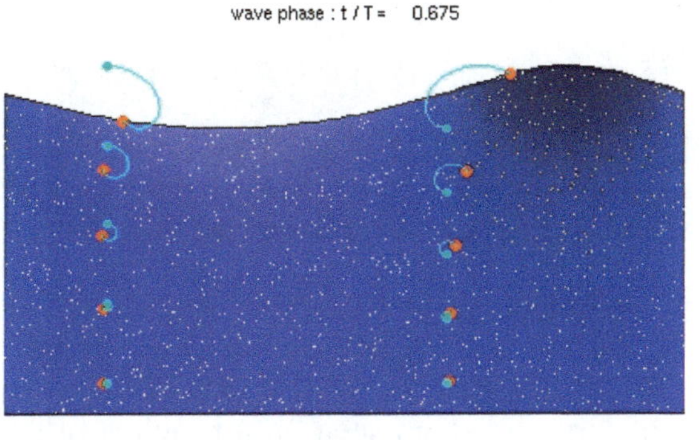

Water particle motion of a deep water wave

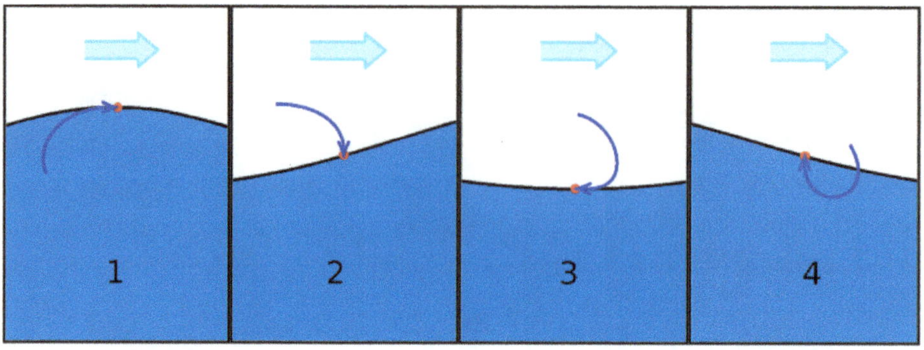

The phases of an ocean surface wave: 1. Wave Crest, where the water masses of the surface layer are moving horizontally in the same direction as the propagating wave front. 2. Falling wave. 3. Trough, where the water masses of the surface layer are moving horizontally in the opposite direction of the wave front direction. 4. Rising wave.

NOAA ship *Delaware II* in bad weather on Georges Bank

The great majority of large breakers seen on a beach result from distant winds. Five factors influence the formation of the flow structures in wind waves:

- Wind speed or strength relative to wave speed—the wind must be moving faster than the wave crest for energy transfer

- The uninterrupted distance of open water over which the wind blows without significant change in direction (called the *fetch*)
- Width of area affected by fetch
- Wind duration—the time over which the wind has blown over a given area
- Water depth

All of these factors work together to determine the size of wind waves and the structures of the flows within:

- Wave height (from high trough to crest)
- Wave length (from crest to crest)
- Wave period (time interval between arrival of consecutive crests at a stationary point)
- Wave propagation direction

A fully developed sea has the maximum wave size theoretically possible for a wind of a specific strength, duration, and fetch. Further exposure to that specific wind could only cause a loss of energy due to the breaking of wave tops and formation of "whitecaps". Waves in a given area typically have a range of heights. For weather reporting and for scientific analysis of wind wave statistics, their characteristic height over a period of time is usually expressed as *significant wave height*. This figure represents an average height of the highest one-third of the waves in a given time period (usually chosen somewhere in the range from 20 minutes to twelve hours), or in a specific wave or storm system. The significant wave height is also the value a "trained observer" (e.g. from a ship's crew) would estimate from visual observation of a sea state. Given the variability of wave height, the largest individual waves are likely to be somewhat less than twice the reported significant wave height for a particular day or storm.

• Sources of wind wave generation: Sea water wave is generated by many kinds of disturbances such as Seismic events, gravity, and crossing wind. The generation of wind wave is initiated by the disturbances of cross wind field on the surface of the sea water. Two major Mechanisms of surface wave formation by winds (a.k.a.'The Miles-Phillips Mechanism') and other sources (ex. earthquakes) of wave formation can explain the generation of wind waves.

However, if one set a flat water surface (Beaufort Point,0) and abrupt cross wind flows on the surface of the water, then the generation of surface wind waves can be explained by following two mechanisms which initiated by normal pressure fluctuations of turbulent winds and parallel wind shear flows.

• The mechanism of the surface wave generation by winds

1) Starts from "Fluctuations of wind" (O. M.Phillips) : the wind wave formation on water surface by wind is started by a random distribution of normal pressure acting on the water from the wind. By the mechanism developed by O. M. Phillips (in 1957), the water surface is initially at rest and wave generation is started by adding turbulent wind flows and then, by the fluctuations of the wind, normal pressure acting on the water surface. From this pressure fluctuation arise normal and tangential stresses to the

surface water, which generates wave behavior on the water surface. It is assumed that:

- The water originally at rest.
- The water is not viscid.
- The water is irrotational.
- There are random distribution of normal pressure to the water surface from the turbulent wind.
- Correlations between air and water motions are neglected.

2) starts from "wind shear forces" on the water surface (J. W. Miles, applied to mainly 2D deep water gravity waves) ; John W. Miles suggested a surface wave generation mechanism which is initiated by turbulent wind shear flows Ua(y), based on the inviscid Orr-Sommerfeld equation in 1957. He found the energy transfer from wind to water surface as a wave speed, c is proportional to the curvature of the velocity profile of wind Ua"(y) at point where the mean wind speed is equal to the wave speed (Ua=c, where, Ua is the Mean turbulent wind speed). Since the wind profile Ua(y) is logarithmic to the water surface, the curvature Ua"(y) have negative sign at the point of Ua=c. This relations show the wind flow transferring its kinetic energy to the water surface at their interface, and arises wave speed, c the growth-rate can be determined by the curvature of the winds ((d^2 Ua)/(dz^2)) at the steering height (Ua (z=z_h)=c) for a given wind speed Ua {*Assumptions*; 1. 2D parallel shear flow, Ua(y) 2. incompressible, inviscid water / wind 3. irrotational water 4. slope of the displacement of surface is small}

Generally these wave formation mechanisms occur together on the ocean surface and arise wind waves and grows up to the fully developed waves.

For example,

If we suppose a very flat sea surface (Beaufort number, 0), and sudden wind flow blows steadily across the sea surface, physical wave generation process will be like;

- Turbulent wind flows form random pressure fluctuations at the sea surface. Small waves with a few centimeters order of wavelengths are generated by the pressure fluctuations. (The Phillips mechanism)
- The cross winds keep acting on the initially fluctuated sea surface causing the waves to become larger. As the waves grow, the pressure differences get larger causing the growth rate to increase. Finally the shear instability expedites the wave growth exponentially. (The Miles mechanism)
- The interactions between the waves on the surface generate longer waves (Hasselmann et al., 1973) and the interaction will transfer wave energy from the shorter waves generated by the Miles mechanism to the waves have slightly lower frequencies than the frequency at the peak wave magnitudes, then finally the waves will be faster than the cross wind speed (Pierson & Moskowitz).

Conditions Necessary for a Fully Developed Sea at Given Wind Speeds, and the Parameters of the Resulting Waves					
Wind Conditions			**Wave Size**		
Wind Speed in One Direction	**Fetch**	**Wind Duration**	**Average Height**	**Average Wavelength**	**Average Period and Speed**
19 km/h (12 mph)	19 km (12 mi)	2 hr	0.27 m (0.89 ft)	8.5 m (28 ft)	3.0 sec 9.3 ft/sec
37 km/h (23 mph)	139 km (86 mi)	10 hr	1.5 m (4.9 ft)	33.8 m (111 ft)	5.7 sec 19.5 ft/sec
56 km/h (35 mph)	518 km (322 mi)	23 hr	4.1 m (13 ft)	76.5 m (251 ft)	8.6 sec 29.2 ft/sec
74 km/h (46 mph)	1,313 km (816 mi)	42 hr	8.5 m (28 ft)	136 m (446 ft)	11.4 sec 39.1 ft/sec
92 km/h (57 mph)	2,627 km (1,632 mi)	69 hr	14.8 m (49 ft)	212.2 m (696 ft)	14.3 sec 48.7 ft/sec

TYPES OF WIND WAVES

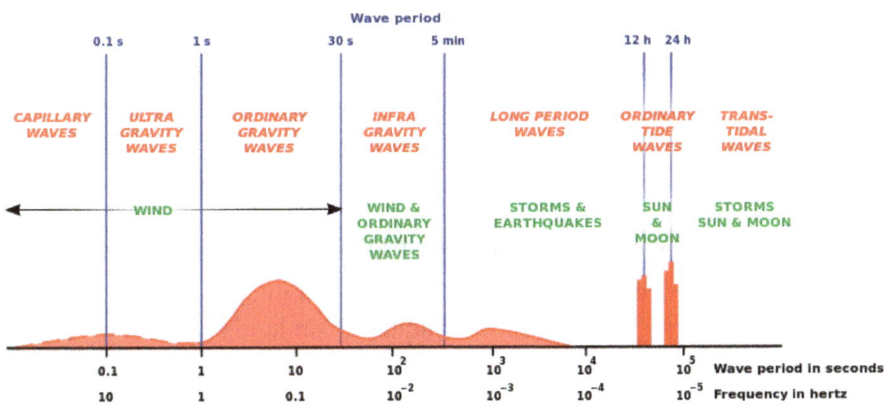

Classification of the spectrum of ocean waves according to wave period.

Surf on a rocky irregular bottom. Porto Covo, west coast of Portugal

Three different types of wind waves develop over time:

- Capillary waves, or ripples
- Seas
- Swells

Ripples appear on smooth water when the wind blows, but will die quickly if the wind stops. The restoring force that allows them to propagate is surface tension. Sea waves are larger-scale, often irregular motions that form under sustained winds. These waves tend to last much longer, even after the wind has died, and the restoring force that allows them to propagate is gravity. As waves propagate away from their area of origin, they naturally separate into groups of common direction and wavelength. The sets of waves formed in this way are known as swells.

Individual "rogue waves" (also called "freak waves", "monster waves", "killer waves", and "king waves") much higher than the other waves in the sea state can occur. In the case of the Draupner wave, its 25 m (82 ft) height was 2.2 times the significant wave height. Such waves are distinct from tides, caused by the Moon and Sun's gravitational pull, tsunamis that are caused by underwater earthquakes or landslides, and waves generated by underwater explosions or the fall of meteorites—all having far longer wavelengths than wind waves.

Yet, the largest ever recorded wind waves are common — not rogue — waves in extreme sea states. For example: 29.1 m (95 ft) high waves have been recorded on the RRS Discovery in a sea with 18.5 m (61 ft) significant wave height, so the highest wave is only 1.6 times the significant wave height. The biggest recorded by a buoy (as of 2011) was 32.3 m (106 ft) high during the 2007 typhoon Krosa near Taiwan.

Ocean waves can be classified based on: the disturbing force(s) that create(s) them; the extent to which the disturbing force(s) continue(s) to influence them after formation; the extent to which the restoring force(s) weaken(s) (or flatten) them; and their wavelength or period. Seismic Sea waves have a period of ~20 minutes, and speeds of 760 km/h (470 mph). Wind waves (deep-water waves) have a period of about 20 seconds.

Wave type	Typical wavelength	Disturbing force	Restoring force
Capillary wave	< 2 cm	Wind	Surface tension
Wind wave	60–150 m (200–490 ft)	Wind over ocean	Gravity
Seiche	Large, variable; a function of basin size	Change in atmospheric pressure, storm surge	Gravity
Seismic sea wave (tsunami)	200 km (120 mi)	Faulting of sea floor, volcanic eruption, landslide	Gravity
Tide	Half the circumference of Earth	Gravitational attraction, rotation of Earth	Gravity

The speed of all ocean waves is controlled by gravity, wavelength, and water depth. Most characteristics of ocean waves depend on the relationship between their wavelength and water depth. Wavelength determines the size of the orbits of water molecules within a wave, but water depth determines the shape of the orbits. The paths of water molecules in a wind wave are circular only when the wave is traveling in deep water. A wave cannot "feel" the bottom when it moves through water deeper than half its wavelength because too little wave energy is contained in the small circles below that depth. Waves moving through water deeper than half their wavelength are known as deep-water waves. On the other hand, the orbits of water molecules in waves moving through shallow water are flattened by the proximity of the sea surface bottom. Waves in water shallower than 1/20 their original wavelength are known as shallow-water waves. Transitional waves travel through water deeper than 1/20 their original wavelength but shallower than half their original wavelength.

- In general, the longer the wavelength, the faster the wave energy will move through the water. For deep-water waves, this relationship is represented with the following formula:

$$C = L/T$$

where C is speed (celerity), L is wavelength, and T is time, or period (in seconds).

The speed of a deep-water wave may also be approximated by:

$$C = \sqrt{gL/2\pi}$$

where g is the acceleration due to gravity, 9.8 meters (32 feet) per second squared. Because g and π (3.14) are constants, the equation can be reduced to:

$$C = 1.251\sqrt{L}$$

when C is measured in meters per second and L in meters. Note that in both formulas the wave speed is proportional to the square root of the wavelength.

The speed of shallow-water waves is described by a different equation that may be written as:

$$C = \sqrt{gd} = 3.1\sqrt{d}$$

where C is speed (in meters per second), g is the acceleration due to gravity, and d is the depth of the water (in meters). The period of a wave remains unchanged regardless of the depth of water through which it is moving. As deep-water waves enter the shallows and feel the bottom, however, their speed is reduced and their crests "bunch up," so their wavelength shortens.

Wave Shoaling and Refraction

As waves travel from deep to shallow water, their shape alters (wave height increases, speed decreases, and length decreases as wave orbits become asymmetrical). This process is called shoaling. Wave refraction is the process by which wave crests realign themselves as a result of decreasing water depths. Varying depths along a wave crest cause the crest to travel at different phase speeds, with those parts of the wave in deeper water moving faster than those in shallow water. This process continues until the crests become (nearly) parallel to the depth contours. Rays—lines normal to wave crests between which a fixed amount of energy flux is contained—converge on local shallows and shoals. Therefore, the wave energy between rays is concentrated as they converge, with a resulting increase in wave height. Because these effects are related to a spatial variation in the phase speed, and because the phase speed also changes with the ambient current – due to the Doppler shift – the same effects of refraction and altering wave height also occur due to current variations. In the case of meeting an adverse current the wave *steepens*, i.e. its wave height increases while the wave length decreases, similar to the shoaling when the water depth decreases.

Wave Breaking

Big wave breaking

Breaking of a wave reaching the beach

Some waves undergo a phenomenon called "breaking". A breaking wave is one whose base can no longer support its top, causing it to collapse. A wave breaks when it runs into shallow water, or when two wave systems oppose and combine forces. When the slope, or steepness ratio, of a wave is too great, breaking is inevitable.

Individual waves in deep water break when the wave steepness—the ratio of the wave height H to the wavelength λ—exceeds about 0.07, so for $H > 0.07\ \lambda$. In shallow water, with the water depth small compared to the wavelength, the individual waves break when their wave height H is larger than 0.8 times the water depth h, that is $H > 0.8\ h$. Waves can also break if the wind grows strong enough to blow the crest off the base of the wave.

Three main types of breaking waves are identified by surfers or surf lifesavers. Their varying characteristics make them more or less suitable for surfing, and present different dangers.

- Spilling, or rolling: these are the safest waves on which to surf. They can be found in most areas with relatively flat shorelines. They are the most common type of shorebreak
- Plunging, or dumping: these break suddenly and can "dump" swimmers—pushing them to the bottom with great force. These are the preferred waves for experienced surfers. Strong offshore winds and long wave periods can cause dumpers. They are often found where there is a sudden rise in the sea floor, such as a reef or sandbar.
- Surging: these may never actually break as they approach the water's edge, as the water below them is very deep. They tend to form on steep shorelines. These waves can knock swimmers over and drag them back into deeper water.

PHYSICS OF WAVES

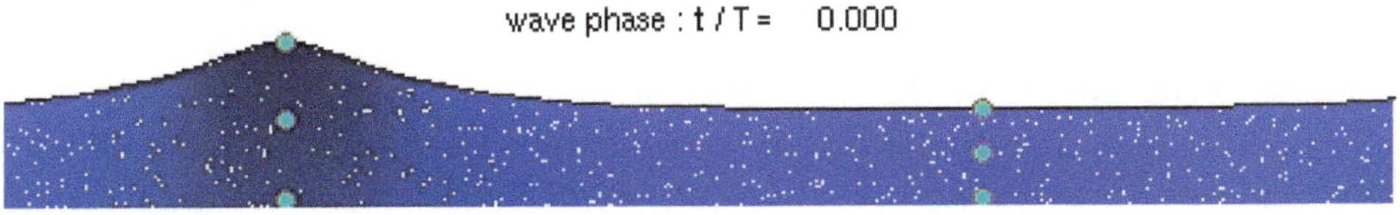

Stokes drift in shallow water waves (Animation)

Wind waves are mechanical waves that propagate. along the interface between water and air; the restoring force is provided by gravity, and so they are often referred to as surface gravity waves. As the wind blows, pressure and friction perturb the equilibrium of the water surface and transfer energy from the air to the water, forming waves. The initial formation of waves by the wind is described in the theory of Phillips from 1957, and the subsequent growth of the small waves has been modeled by Miles, also in 1957.

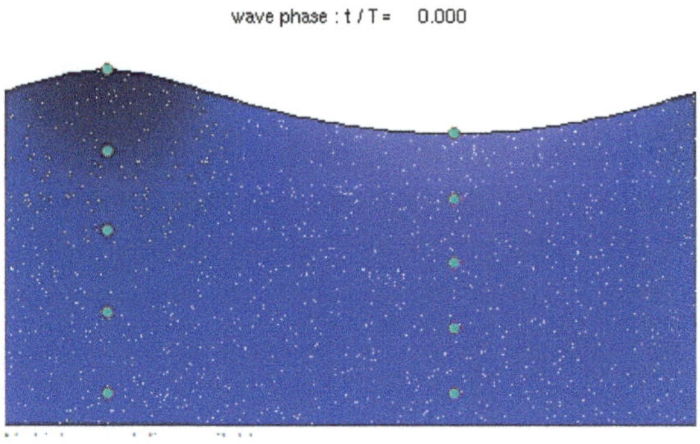

Stokes drift in a deeper water wave (Animation)

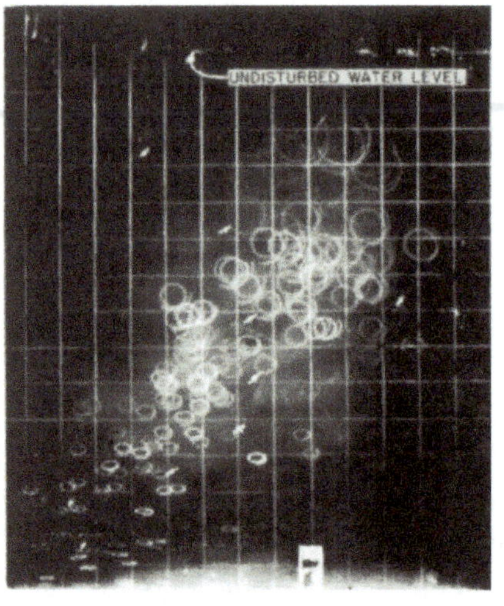

Photograph of the water particle orbits under a – progressive and periodic – surface gravity wave in a wave flume. The wave conditions are: mean water depth d = 2.50 ft (0.76 m), wave height H = 0.339 ft (0.103 m), wavelength λ = 6.42 ft (1.96 m), period$T$ = 1.12 s.

In linear plane waves of one wavelength in deep water, parcels near the surface move not plainly up and down but in circular orbits: forward above and backward below (compared the wave propagation direction). As a result, the surface of the water forms not an exact sine wave, but more a trochoid with the sharper curves upwards—as modeled in trochoidal wave theory.

When waves propagate in shallow water, (where the depth is less than half the wavelength) the particle trajectories are compressed into ellipses.

In reality, for finite values of the wave amplitude (height), the particle paths do not form closed orbits; rather, after the passage of each crest, particles are displaced slightly from their previous positions, a phenomenon known as Stokes drift.

As the depth below the free surface increases, the radius of the circular motion decreases. At a depth equal to half the wavelength λ, the orbital movement has decayed to less than 5% of its value at the surface. The phase speed (also called the celerity) of a surface gravity wave is – for pure periodic wave motion of small-amplitude waves – well approximated by

$$c = \sqrt{\frac{g\lambda}{2\pi} \tanh\left(\frac{2\pi d}{\lambda}\right)}$$

where

c = phase speed;

λ = wavelength;

d = water depth;

g = acceleration due to gravity at the Earth's surface.

In deep water, where $d \geq \frac{1}{2}\lambda$, so $\frac{2\pi d}{\lambda} \geq \pi$ and the hyperbolic tangent approaches 1, the speed c approximates

$$c_{\text{deep}} = \sqrt{\frac{g\lambda}{2\pi}}.$$

In SI units, with c_{deep} in m/s, $c_{\text{deep}} \approx 1.25\sqrt{\lambda}$, when λ is measured in metres. This expression tells us that waves of different wavelengths travel at different speeds. The fastest waves in a storm are the ones with the longest wavelength. As a result, after a storm, the first waves to arrive on the coast are the long-wavelength swells.

For intermediate and shallow water, the Boussinesq equations are applicable, combining frequency dispersion and nonlinear effects. And in very shallow water, the shallow water equations can be used.

If the wavelength is very long compared to the water depth, the phase speed (by taking the limit of c when the wavelength approaches infinity) can be approximated by

$$c_{\text{shallow}} = \lim_{\lambda \to \infty} c = \sqrt{gd}.$$

On the other hand, for very short wavelengths, surface tension plays an important role and the phase speed of these gravity-capillary waves can (in deep water) be approximated by

$$c_{\text{gravity-capillary}} = \sqrt{\frac{g\lambda}{2\pi} + \frac{2\pi S}{\rho\lambda}}$$

where

S = surface tension of the air-water interface;

ρ = density of the water.

When several wave trains are present, as is always the case in nature, the waves form groups. In deep water the groups travel at a group velocity which is half of the phase speed. Following a single wave in a group one can see the wave appearing at the back of the group, growing and finally disappearing at the front of the group.

As the water depth p decreases towards the coast, this will have an effect: wave height changes due to wave shoaling and refraction. As the wave height increases, the wave may become unstable when the crest of the wave moves faster than the trough. This causes *surf*, a breaking of the waves.

The movement of wind waves can be captured by wave energy devices. The energy density (per unit area) of regular sinusoidal waves depends on the water density, ρ gravity acceleration g and the wave height H (which, for regular waves, is equal to twice the amplitude, a):

$$E = \frac{1}{8}\rho g H^2 = \frac{1}{2}\rho g a^2.$$

The velocity of propagation of this energy is the group velocity.

Wind Wave Models

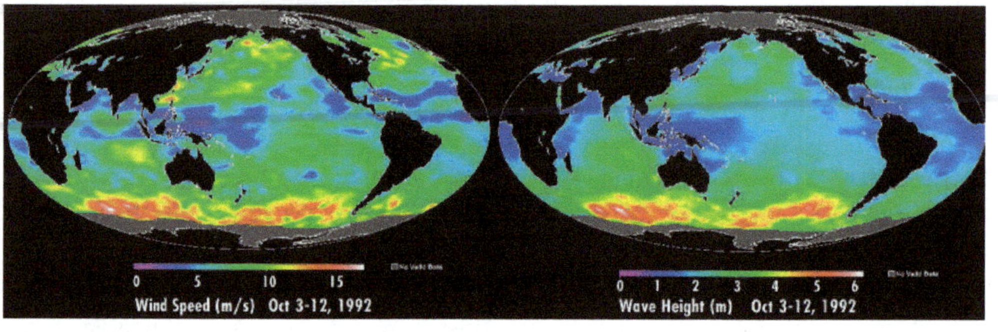

The image shows the global distribution of wind speed and wave height as observed by NASA's TOPEX/Poseidon's dual-frequency radar altimeter from October 3 to October 12, 1992. Simultaneous observations of wind speed and wave height are helping scientists to predict ocean waves. Wind speed is determined by the strength of the radar signal after it has bounced off the ocean surface and returned to the satellite. A calm sea serves as a good reflector and returns a strong signal; a rough sea tends to scatter the signals and returns a weak pulse. Wave height is determined by the shape of the return radar pulse. A calm sea with low waves returns a condensed pulse whereas a rough sea with high waves returns a stretched pulse. Comparing the two images above shows a high degree of correlation between wind speed and wave height. The strongest winds (33.6 mph; 54.1 km/h) and highest waves are found in the Southern Ocean. The weakest winds—shown as areas of magenta and dark blue—are generally found in the tropical Oceans.

Surfers are very interested in the wave forecasts. There are many websites that provide predictions of the surf quality for the upcoming days and weeks. Wind wave models are driven by more general weather models that predict the winds and pressures over the oceans, seas and lakes.

Wind wave models are also an important part of examining the impact of shore protection and beach nourishment proposals. For many beach areas there is only patchy information about the wave climate, therefore estimating the effect of wind waves is important for managing littoral environments.

Seismic Signals

Ocean water waves generate land seismic waves that propagate hundreds of kilometers into the land. These seismic signals usually have the period of 6 ± 2 seconds. Such recordings were first reported and understood in about 1900.

There are two types of seismic "ocean waves". The primary waves are generated in shallow waters by direct water wave-land interaction and have the same period as the water waves (10 to 16 seconds). The more powerful secondary waves are generated by the superposition of ocean waves of equal period traveling in opposite directions, thus generating standing gravity waves – with an associated pressure oscillation at half the period, which is not diminishing with depth. The theory for microseism generation by standing waves was provided by Michael Longuet-Higgins in 1950, after in 1941 Pierre Bernard suggested this relation with standing waves on the basis of observations.

INTERNAL WAVES

Internal waves can form at the boundary between water layers of different densities. These sub-surface waves are called internal waves. As is the case with ocean waves at the air-ocean interface, internal waves possess troughs, crests, wavelength, and period. Internal waves move very slowly because the density difference between the joined media is very small. Internal waves occur in the ocean at the base of the pycnocline, especially at the bottom edge of a steep thermocline. The wave height of internal waves may be greater than 30 meters (98 feet), causing the pycnocline to undulate slowly through a considerable depth. Their wavelength often exceeds 0.8 kilometres (0.50 mi) and their periods are typically 5 to 8 minutes. Internal waves are generated by wind energy, tidal energy, and ocean currents. Surface manifestations of internal waves have been photographed from space.

Internal waves may mix nutrients into surface water and trigger plankton blooms. They can also affect submarines and oil platforms.

REFERENCES

1. Tolman, H. L. (23 June 2010). Mahmood, M.F., ed. "CBMS Conference Proceedings on Water Waves: Theory and Experiment" (PDF). Howard University, US, 13–18 May 2008: World Scientific Publications. ISBN 978-981-4304-23-8.
2. Holthuijsen (2007), page 5.
3. Lorenz, R. D. and A. G. Hayes, "The Growth of Wind-Waves in Titan's

Hydrocarbon Seas", Icarus, 219, 468–475, 2012

4. Young, I. R. (1999). Wind generated ocean waves. Elsevier. ISBN 0-08-043317-0.p. 83.
5. Weisse, Ralf; von Storch, Hans (2008). Marine climate change: Ocean waves, storms and surges in the perspective of climate change. Springer. p. 51. ISBN 978-3-540-25316-7.
6. Phillips, O. M. (1957), "On the generation of waves by turbulent wind", Journal of Fluid Mechanics 2 (5): 417–445, Bibcode:1957JFM.....2..417P, doi:10.1017/S0022112057000233
7. Miles, J. W. (1957), "On the generation of surface waves by shear flows", Journal of Fluid Mechanics 3 (2): 185–204, Bibcode:1957JFM.....3..185M, doi:10.1017/S0022112057000567
8. Chapter 16, Ocean Waves
9. Hasselmann K., T. P. Barnett, E. Bouws, H. Carlson, D. E. Cartwright, K. Enke, J. A. Ewing, H. Gienapp, D. E. Hasselmann, P. Kruseman, A. Meerburg, P. Mller, D. J. Olbers, K. Richter, W. Sell, and H. Walden. "Measurements of wind-wave growth and swell decay during the Joint North Sea Wave Project (JONSWAP)" 'Ergnzungsheft zur Deutschen Hydrographischen Zeitschrift Reihe, A(8) (Nr. 12), p.95, 1973.
10. Pierson, Willard J., Jr. and Moskowitz, Lionel A. "Proposed Spectral Form for Fully Developed Wind Seas Based on the Similarity Theory of S. A. Kitaigorodskii", Journal of Geophysical Research, Vol. 69, p.5181-5190, 1964.
11. Munk, Walter H. (1950). "Proceedings 1st International Conference on Coastal Engineering". Long Beach, California: ASCE: 1–4.
12. Holliday, Naomi P.; Yelland, Margaret J.; Pascal, Robin; Swail, Val R.; Taylor, Peter K.; Griffiths, Colin R.; Kent, Elizabeth (2006). "Were extreme waves in the Rockall Trough the largest ever recorded?". Geophysical Research Letters. 33 (L05613). Bibcode:2006GeoRL..3305613H. doi:10.1029/2005GL025238.
13. P. C. Liu; H. S. Chen; D.-J. Doong; C. C. Kao; Y.-J. G. Hsu (11 June 2008). "Monstrous ocean waves during typhoon Krosa" (PDF). Annales Geophysicae. European Geosciences Union. 26: 1327–1329. Bibcode:2008AnGeo..26.1327L. doi:10.5194/angeo-26-1327-2008.
14. Tom Garrison (2009). Oceanography: An Invitation to Marine Science (7th Edition). Yolanda Cossio. ISBN 978-0495391937.
15. Longuet-Higgins, M. S.; Stewart, R. W. (1964). "Radiation stresses in water waves; a physical discussion, with applications". Deep-Sea Research. 11 (4): 529–562. Bibcode:1964DSROA..11..529L. doi:10.1016/0011-7471(64)90001-4.
16. R.J. Dean and R.A. Dalrymple (2002). Coastal processes with engineering applications. Cambridge University Press. ISBN 0-521-60275-0. p. 96–97.
17. Phillips, O. M. (1957). "On the generation of waves by turbulent wind". Journal of Fluid Mechanics. 2 (5): 417–445. Bibcode:1957JFM.....2..417P. doi:10.1017/S0022112057000233.

18. Miles, J. W. (1957). "On the generation of surface waves by shear flows". Journal of Fluid Mechanics. 3 (2): 185–204. Bibcode:1957JFM.....3..185M. doi:10.1017/S0022112057000567.
19. Figure 6 from: Wiegel, R. L.; Johnson, J. W. (1950). "Proceedings 1st International Conference on Coastal Engineering". Long Beach, California: ASCE: 5–21.
20. For the particle trajectories within the framework of linear wave theory, see for instance:
21. Phillips (1977), page 44.
22. Lamb, H. (1994). Hydrodynamics (6th ed.). Cambridge University Press. ISBN 978-0-521-45868-9. Originally published in 1879, the 6th extended edition appeared first in 1932. See §229, page 367.
23. L. D. Landau and E. M. Lifshitz (1986). Fluid mechanics. Course of Theoretical Physics. 6 (Second revised ed.). Pergamon Press. ISBN 0-08-033932-8. See page 33.
24. A good illustration of the wave motion according to linear theory is given by Prof. Robert Dalrymple's Java applet.
25. For nonlinear waves, the particle paths are not closed, as found by George Gabriel Stokes in 1847, see the original paper by Stokes. Or in Phillips (1977), page 44: "To this order, it is evident that the particle paths are not exactly closed ... pointed out by Stokes (1847) in his classical investigation".
26. Solutions of the particle trajectories in fully nonlinear periodic waves and the Lagrangian wave period they experience can for instance be found in:
27. J. M. Williams (1981). "Limiting gravity waves in water of finite depth". Philosophical Transactions of the Royal Society A. 302 (1466): 139–188. Bibcode:1981RSPTA.302..139W. doi:10.1098/rsta.1981.0159.
28. J. M. Williams (1985). Tables of progressive gravity waves. Pitman. ISBN 978-0-273-08733-5.
29. Carl Nordling, Jonny Östermalm (2006). Physics Handbook for Science and Engineering (Eight ed.). Studentliteratur. p. 263. ISBN 978-91-44-04453-8.
30. In deep water, the group velocity is half the phase velocity, as is shown here. Another reference is .
31. Peter Bormann. Seismic Signals and Noise
32. Bernard, P. (1941). "Sur certaines proprietes de la boule etudiees a l›aide des enregistrements seismographiques". Bulletin de l›Institut océanographique de Monaco. 800: 1–19.
33. Longuet-Higgins, M. S. (1950). "A theory of the origin of microseisms". Philosophical Transactions of the Royal Society A. 243 (857): 1–35. Bibcode:1950RSPTA.243....1L. doi:10.1098/rsta.1950.0012.

CHAPTER 8

Coastal Landforms and Processes

Chapter Contents

Waves, tides, and currents help create coastlines. When waves crash onto shore, they wear away at, or erode, the land. But they also leave behind little parts of the sea, such as shells, sand dollars, seaweeds, and hermit crabs. Sometimes these objects end up as more permanent parts of the coastline.

INTRODUCTION

A coastline or a seashore is the area where land meets the sea or ocean, or a line that forms the boundary between the land and the ocean or a lake. A precise line that can be called a coastline cannot be determined due to the Coastline paradox.

The term coastal zone is a region where interaction of the sea and land processes occurs. Both the terms coast and coastal are often used to describe a geographic location or region; for example, New Zealand's West Coast, or the East and West Coasts of the United States. Edinburgh for example is a city on the coast of Scotland.

A pelagic coast refers to a coast which fronts the open ocean, as opposed to a more sheltered coast in a gulf or bay. A shore, on the other hand, can refer to parts of the land which adjoin any large body of water, including oceans (sea shore) and lakes (lake shore). Similarly, the somewhat related term refers to the land alongside or sloping down to a river (riverbank) or to a body of water smaller than a lake. "Bank" is also used in some parts of the world to refer to an artificial ridge of earth intended to retain the water of a river or pond; in other places this may be called a levee.

While many scientific experts might agree on a common definition of the term "coast", the delineation of the extents of a coast differ according to jurisdiction, with many scientific and government authorities in various countries differing for economic and social policy reasons. According to the UN atlas, 44% of people live within 150 kilometres (93 miles) of the sea.

Rugged coastline of the West Coast Region of New Zealand

Formation

Atlantic rocky coastline, showing a surf area. Porto Covo, west coast of Portugal

Tidesoften determine the range over whichsedimentis deposited or eroded. Areas with high tidal ranges allow waves to reach farther up the shore, and areas with lower tidal ranges produce deprossosition at a smaller elevation interval. The tidal range is influenced by the size and shape of the coastline. Tides do not typically cause erosion by themselves; however,tidal borescan erode as the waves surge up riverestuariesfrom the ocean. Waves erode coastline as they break on shore releasing their energy; the larger the wave the more energy it releases and the more sediment it moves. Coastlines with longer shores have more room for the waves to disperse their energy, while coasts with cliffs and short shore faces give little room for the wave energy to be dispersed. In these areas the wave energy breaking against the cliffs is higher, and air and water are compressed into cracks in the rock, forcing the rock apart, breaking it down. Sediment deposited by waves comes from eroded cliff faces and is moved along the coastline by the waves. This forms anabrasionorcliffed coast.

Sediment deposited by rivers is the dominant influence on the amount of sediment located on a coastline.Today riverine deposition at the coast is often blocked by dams and other human regulatory devices, which remove the sediment from the stream by causing it to be deposited inland.

Like the ocean which shapes them, coasts are a dynamic environment with constant change. The Earth's natural processes, particularlysea level rises, waves and variousweatherphenomena, have resulted in theerosion,accretionand reshaping of coasts as well as flooding and creation ofcontinental shelvesand drowned river valleys (rias).

ENVIRONMENTAL IMPORTANCE

Somaliahas the longest coastline in Africa

The coast and its adjacent areas on and off shore are an important part of a localecosystem: the mixture offresh waterandsalt water(brackish water) inestuariesprovides many nutrients for marine life.Salt marshesandbeachesalso support a diversity ofplants,a nimalsandinsectscrucial to thefood chain.

The high level ofbiodiversitycreates a high level of biological activity, which has attracted human activity for thousands of years.

HUMAN IMPACTS

Human Uses of Coasts

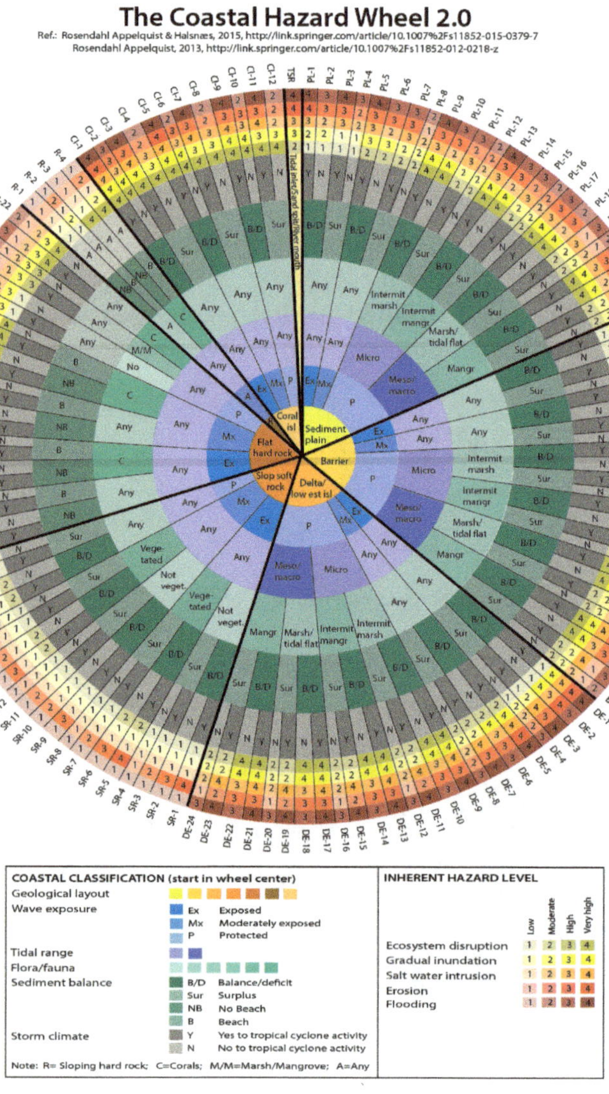

The Coastal Hazard Wheel system published by UNEP for global coastal management

A settled coastline inMarblehead, Massachusetts. Once a fishing port, the harbor is now dedicated to tourism and pleasure boating. Observe that the sand and rocks have been darkened by oil slick up to the high-water line.

This stretch of coast in Tanzania's capital Dar es Salaam serves as a public waste dump.

Housesclose to the coast, like these inTiburon, California, may be especially desirable properties.

View of sea coast from top of a hill atVisakhapatnaminIndia

More and more of the world's people live in coastal regions.Many major cities are on or near goodharborsand haveportfacilities. Somelandlockedplaces have achieved port status by buildingcanals.

The coast is a frontier that nations have typically defended against military invaders, smugglers and illegal migrants. Fixedcoastal defenseshave long been erected in many nations and coastal countries typically have anavyand some form ofcoast guard.

Coasts, especially those withbeachesand warm water, attracttourists. In manyisland nationssuch as those of the Mediterranean, South Pacific and Caribbean, tourism is central to the economy. Coasts offer recreational activities such asswimming,fishing,surfing,boati ng, andsunbathing.Growth managementcan be a challenge for coastallocal authoritieswho often struggle to provide theinfrastructurerequired by new residents.

Threats to a Coast

Coasts also face manyhuman-induced environmental impacts. The human influence onclimate changeis thought to contribute to an accelerated trend insea level risewhich threatens coastalhabitats.

Pollutioncan occur from a number of sources:garbage and industrial debris; thetransportation of petroleumintankers, increasing the probability of largeoil spills; small oil spills created by large and small vessels, which flushbilgewater into the ocean.

Fishing has declined due tohabitat degradation,overfishing,trawling,bycatchandclimate change. Since the growth of global fishing enterprises after the 1950s, intensive fishing has spread from a few concentrated areas to encompass nearly all fisheries. The scraping of the ocean floor inbottom draggingis devastating tocoral,spongesand other long-lived species that do not recover quickly. This destruction alters the functioning of the ecosystem and can permanently alter species composition and biodiversity.Bycatch, the capture of unintended species in the course of fishing, is typically returned to the ocean only to die from injuries or exposure. Bycatch represents about a quarter of all marine catch. In the case ofshrimpcapture, the bycatch is five times larger than the shrimp caught.

It is believed that meltingArctic icewill cause sea levels to rise and flood coastal areas.

Conservation

Extraordinarypopulation growthin the 21st century has placed stress on the planet's ecosystems. For example, onSaint Lucia, harvestingmangrovefor timber and clearing for fishing reduced the mangrove forests, resulting in aloss of habitatand spawning grounds for marine life that was unique to the area. These forests also helped to stabilize the coastline. Conservation efforts since the 1980s have partiallyrestored the ecosystem.

TYPES OF COAST

According to one principle of classification, an emergent coastline is a coastline which has experienced a fall in sea level, because of either a global sea level change, or local uplift. Emergent coastlines are identifiable by the coastallandforms, which are above the high tide mark, such asraised beaches. In contrast, a submergent coastline is one where the sea level has risen, due to a global sea level change, localsubsidence, orisostatic rebound. Submergent coastlines are identifiable by their submerged, or "drowned" landforms, such asrias(drowned valleys) andfjords.

According to a second principle of classification, a concordant coastline is a coastline where bands of different rock types run parallel to the shore. These rock types are usually of varyingresistance, so the coastline forms distinctive landforms, such as coves. Discordant coastlines feature distinctive landforms because the rocks areerodedby oceanwaves. The less resistant rocks erode faster, creatinginletsorbay; the more resistant rocks erode more slowly, remaining asheadlandsoroutcroppings.

Other coastal categories:

- Acliffed coastor abrasion coast is one where marine action has produced steep declivities known ascliffs.
- Aflat coastis one where the land gradually descends into the sea.
- Agraded shorelineis one where wind and water action has produced a flat and straight coastline.

Coastal Landforms

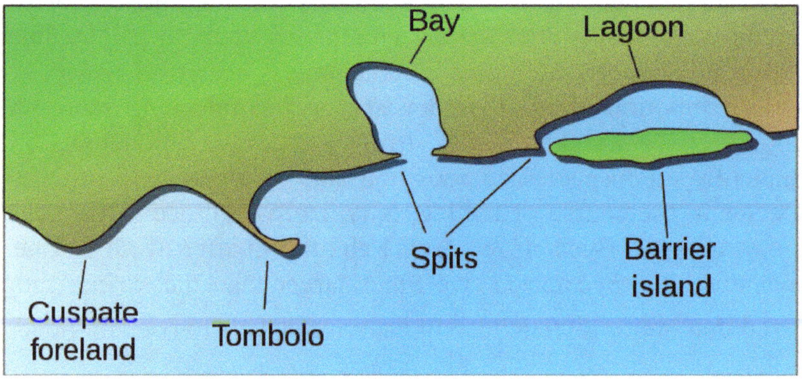

Coastal landforms. The feature shown here as abaywould, in certain (mainly southern) parts ofBritain, be called acove. That between the cuspate foreland and the tombolo is a British bay.

The following articles describe some coastal landforms

• Bay	• Cove
• Headland	• Peninsula

Cliff Erosion

- Much of the sediment deposited along a coast is the result of erosion of a surroundingcliff, or bluff.Sea cliffsretreat landward because of the constant undercutting of slopes by waves. If the slope/cliff being undercut is made of unconsolidated sediment it will erode at a much faster rate then a cliff made of bedrock.
- Anatural archis formed when a sea stack is eroded through by waves.

- Sea cavesare made when certain rock beds are more susceptible to erosion than the surrounding rock beds because of different areas of weakness. These areas are eroded at a faster pace creating a hole or crevasse that, through time, by means of wave action and erosion, becomes a cave.
- Astackis formed when a headland is eroded away by wave and wind action.
- Astumpis a shortened sea stack that has been eroded away or fallen because of instability.
- Wave-cut notches are caused by the undercutting of overhanging slopes which leads to increased stress on cliff material and a greater probability that the slope material will fall. The fallen debris accumulates at the bottom of the cliff and is eventually removed by waves.
- Awave-cut platformforms after erosion and retreat of a sea cliff has been occurring for a long time. Gently sloping wave-cut platforms develop early on in the first stages of cliff retreat. Later the length of the platform decreases because the waves lose their energy as they break further off shore.

Coastal features formed by sediment

Beach	Cuspate foreland	Ria	Spit
Beach cusps	Dune system	Shoal	Surge channel
	Mudflat	Strand plain	Tombolo
	Raised beach		

Coastal features formed by another feature

- Lagoon
- Salt marsh

Other features on the coast

• Concordant coastline	• Fjord	• Island arc
• Discordant coastline	• Island	• Machair

Coastal processes

The following articles describe the various geologic processes that affect a coastal zone:

Attrition Currents Denudation Deposition Erosion Flooding	Longshore drift Saltation Sea level change eustatic isostatic	Sedimentation Coastal sediment supply sediment transport solution subaerialprocesses suspension	Tides Water waves diffraction refraction wave breaking wave shoaling Weathering

WILDLIFE

Animals

Some of the animals live along a typical coast. There are animals likepuffins,sea turtlesandrockhopper penguins.Sea snailsand various kinds ofbarnacleslive on the coast and scavenge on food deposited by the sea. Most coastal animals are used to humans in developed areas, such asdolphinsandseagullswho eat food thrown for them by tourists. Since the coastal areas are all part of thelittoral zone, there is a profusion of marine life found just off-coast.

There are many kinds ofseabirdson the coast.Pelicansandcormorantsjoin up withternsandoystercatchersto forage for fish and shellfish on the coast. There are sea lions on the coast of Wales and other countries.

Plants

Coastal areas are famous for theirkelpbeds. Kelp is a fast-growingseaweedthat grows up to a metre a day.Coralsandsea anemonesare true animals, but live a lifestyle similar to that of plants.Mangroves,seagrassesandsalt marshare important coastal vegetation types in tropical and temperate environments respectively.

COASTLINE STATISTICS

Coastline Problem

Shortly before 1951,Lewis Fry Richardson, in researching the possible effect of border lengths on the probability of war, noticed that thePortuguesereported their measured border withSpainto be 987km, but the Spanish reported it as 1214km. This was the beginning of the coastline problem, which is a mathematical uncertainty inherent in the measurement of boundaries that are irregular.

The prevailing method of estimating the length of a border (or coastline) was to lay out*n*equal straight-line segments of length*l*withdividerson a map or aerial photograph. Each end of the segment must be on the boundary. Investigating the discrepancies in border estimation, Richardson discovered what is now termed the**Richardson Effect**: the sum of the segments is inversely proportional to the common length of the segments. In effect, the shorter the ruler, the longer the measured border; the Spanish and Portuguese geographers were simply using different-length rulers.

The result most astounding to Richardson is that, under certain circumstances, as*l*approaches zero, the length of the coastline approachesinfinity. Richardson had believed, based on Euclidean geometry, that a coastline would approach a fixed length, as do similar estimations of regular geometric figures. For example, theperimeterof a regularpolygoninscribed in acircleapproaches thecircumferencewith increasing numbers of sides (and decrease in the length of one side). Ingeometric measure theorysuch a smooth curve as the circle that can be approximated by small straight segments with a definite limit is termed arectifiable curve.

Measuring a Coastline

More than a decade after Richardson completed his work,Benoit Mandelbrotdeveloped a new branch ofmathematics,fractal geometry, to describe just such non-rectifiable complexes in nature as the infinite coastline.His own definition of the new figure serving as the basis for his study is:

I coined*fractal*from theLatinadjective*fractus*. The corresponding Latin verb*frangere*means "to break:" to create irregular fragments. It is therefore sensible ... that, in addition to "fragmented" ...*fractus*should also mean "irregular."

A key property of the fractal isself-similarity; that is, at anyscalethe same general configuration appears. A coastline is perceived as bays alternating with promontories. In the hypothetical situation that a given coastline has this property of self-similarity, then no matter how greatly any one small section of coastline is magnified, a similar pattern of smaller bays and promontories superimposed on larger bays and promontories appears, right down to the grains of sand. At that scale the coastline appears as a momentarily shifting, potentially infinitely long thread with astochasticarrangement of bays and promontories formed from the small objects at hand. In such an environment (as opposed to smooth curves) Mandelbrot assertscoastline length turns out to be an elusive notion that slips between the fingers of those who want to grasp it.

There are different kinds of fractals. A coastline with the stated property is in a first category of fractals, namely curves whose fractal dimension is greater than 1. That last statement represents an extension by Mandelbrot of Richardson›s thought. Mandelbrot›s statement of the Richardson Effect is:

$$L(\epsilon) \sim F\epsilon^{1-D}$$

where L, coastline length, a function of the measurement unit, ε, is approximated by the expression. F is a constant and D is a parameter that Richardson found depended on the coastline approximated by L. He gave no theoretical explanation but Mandelbrot identified D with a non-integer form of the Hausdorff dimension, later the fractal dimension. Rearranging the right side of the expression obtains:

$$\frac{F}{\epsilon^D} \cdot \epsilon$$

where $F\varepsilon^{-D}$must be the number of units ε required to obtain L. The fractal dimension is the number of the dimensions of the figure being used to approximate the fractal: 0 for a dot, 1 for a line, 2 for a square. D in the expression is between 1 and 2, for coastlines typically less than 1.5. The broken line measuring the coast does not extend in one direction nor does it represent an area, but is intermediate. It can be interpreted as a thick line or band of width 2ε. More broken coastlines have greater D and therefore L is longer for the same ε. Mandelbrot showed that D is independent of ε.

REFERENCES

1. Burke, Lauretta A.; Kura, Yumiko; Kassem, Ken; Revenga, Carmen; Spalding, Mark; McAllister, Don (2001). Hutter, Carolynne, ed. "Pilot Analysis of Global Ecosystems". World Resources Institute. ISBN 978-1-56973-458-2. |contribution= ignored (help)
2. Davidson, Jon P.; Reed, Walter E.; Davis, Paul M. (2002). Exploring Earth: An Introduction to Physical Geology. Upper Saddle River, NJ: Prentice-Hall Inc. ISBN 978-0-13-018372-9.
3. Easterbrook, Don J. (1999). Surface Processes and Landforms (2 ed.). Upper Saddle River, NJ: Prentice-Hall Inc. ISBN 978-0-13-860958-0.
4. Haslett, Simon K. (2009). Coastal Systems (2nd Edition). introduction to environment. New York: Routledge. ISBN 978-0-415-44060-8.
5. Mandelbrot, Benoit B. "II.5 How long is the coast of Britain?". The Fractal Geometry of Nature. Macmillan. pp. 25–33. ISBN 978-0-7167-1186-5.

CHAPTER 9

Marine Ecosystem and Conservation

Chapter Contents

- **Introduction**
- **Overview**
- **Human Impact**
- **Laws and Treaties**
- **Extinct and Endangered Species**
- **History of Marine Resources Conservation**

As the official archive for global oceanographic data, the National Oceanographic Data Center (NODC) provides data access to researchers, resource managers, and policy makers who must evaluate our natural resources and determine conservation needs. Data on ocean currents, temperature and salinity are used to study migration routes of fish populations, whales, turtles, angel fishand even an occasional wayward Florida manatee. Plankton and chlorophyll data are available to study trends of ocean productivity. NODC is also deeply involved in the establishment of a coral reef information system known as CoRIS, which will integrate reef monitoring data and the chemical, biological and physical data of the surrounding areas. This system will help researchers to better understand, and perhaps help mitigate, the depletion the world's coral reefs.

INTRODUCTION

Marine resources conservation also known as marine conservation, is the protection and preservation of ecosystems in oceans and seas. Marine resources conservation focuses on limiting human-caused damage to marine ecosystems, restoring damaged marine ecosystems, and preserving vulnerable marine species.

Coral reefs have a great amount of biodiversity.

OVERVIEW

Marine resources conservation is a response to biological issues such as extinction and marine habitats change. Marine conservation is the study of conserving physical and biological marine resources and ecosystem functions. This is a relatively new discipline. Marine conservationists rely on a combination of scientific principles derived from marine biology, oceanography, and fisheries science, as well as on human factors such as demand for marine resources and marine law, economics and policy in order to determine how to best protect and conserve marine species and ecosystems. Marine resources conservation can be seen as sub-discipline of conservation biology.

Coral Reefs

Coral reefs are the epicenter of immense amounts of biodiversity, and are a key player in the survival of an entire ecosystem. They provide various marine animals with food, protection, and shelter which keep generations of species alive. Furthermore, coral reefs are an integral part of sustaining human life through serving as a food source (i.e., fish and mollusks) as well as a marine space for ecotourism which provides economic benefits. Also, humans are now conducting research regarding the use of corals as new potential sources for pharmaceuticals (i.e. steroids and anti-inflammatory drugs).

Unfortunately, because of the human impact on coral reefs, these ecosystems are becoming increasingly degraded and in need of conservation. The biggest threats include overfishing, destructive fishing practices, and sedimentation and pollution from land-based sources. This, in conjunction with increased carbon in oceans, coral bleaching, and diseases, means that there are no pristine reefs anywhere in the world. Up to 88%

of coral reefs in Southeast Asia are now threatened, with 50% of those reefs at either "high" or "very high" risk of disappearing, which directly affects the biodiversity and survival of species dependent on coral.

This is especially harmful to island nations such as Samoa, Indonesia, and the Philippines, because many people there depend on the coral reef ecosystems to feed their families and to make a living. However, many fishermen are unable to catch as many fish as they used to, so they are increasingly using cyanide and dynamite in fishing, which further degrades the coral reef ecosystem. This perpetuation of bad habits simply leads to the further decline of coral reefs and therefore perpetuates the problem. One way of stopping this cycle is by educating the local community about why the conservation of marine spaces that include coral reefs is important. Once the local communities understand the personal stakes, then they will fight to preserve the reefs. Conserving coral reefs has many economic, social, and ecological benefits, not only for the people who live on these islands, but for people throughout the world.

HUMAN IMPACT

The deterioration of coral reefs is mainly linked to human activities – 88% of reefs are threatened through various reasons as listed above, including excessive amounts of CO_2 (carbon dioxide) emissions. Oceans absorb approximately 1/3 of the CO_2 produced by humans, which has detrimental effects on the marine environment. The increasing levels of CO_2 in oceans change the seawater chemistry by decreasing the pH. This process is also known as ocean acidification. Acidification negatively affects the carbonate buffering system and drops the carbonate saturation by 30%, which results in a decrease in reef calcification.Reductions in calcification have negative implications on calcifiers, such as corals and shellfish. Some examples include diminishing coral resilience from bleaching, decreasing organisms› ability to fight off predators, inhibiting their potential to compete for food, and altering behavior patterns. When the bottom of the food web declines tremendously due to acidification, the food web and the whole marine conservation effort is jeopardized. Although humans cause the greatest threat to the marine environment, they also have the ability to create effective management plans that will be the key to successful marine conservation. Although the most widely known conservation tool is the MPA, one of the best marine conservation tools simply stems from smarter individual choices made in efforts to reduce CO_2emissions on a daily basis.

Techniques

Strategies and techniques for marine conservation tend to combine theoretical disciplines, such as population biology, with practical conservation strategies, such as setting up protected areas, as with marine protected areas (MPAs) or Voluntary Marine Conservation Areas. Other techniques include developing sustainable fisheries and restoring the populations of endangered species through artificial means.

Another focus of conservationists is on curtailing human activities that are detrimental to either marine ecosystems or species through policy, techniques such as fishing quotas, like those set up by the Northwest Atlantic Fisheries Organization, or laws such as those listed below. Recognizing the economics involved in human use of marine ecosystems is key, as is education of the public about conservation issues. This includes educating tourists that come to an area who might not be familiar with certain regulations regarding the marine habitat. One example of this is a project called Green Fins based in Southeast Asia that uses the scuba diving industry to educate the public. This project, implemented by UNEP, encourages scuba diving operators to educate their students about the importance of marine conservation and encourage them to dive in an environmentally friendly manner that does not damage coral reefs or associated marine ecosystems.

Technology and Halfway Technology

Marine conservation technologies are used to protect endangered and threatened marine organisms and/or habitat. These technologies are innovative and revolutionary because they reduce bycatch, increase the survivorship and health of marine life and habitat, and benefit fishermen who depend on the resources for profit. Examples of technologies include marine protected areas (MPAs), turtle excluder devices (TEDs), autonomous recording unit, pop-up satellite archival tag, and radio-frequency identification (RFID). Commercial practicality plays an important role in the success of marine conservation because it is necessary to cater to the needs of fishermen while also protecting marine life.

Pop-up satellite archival tag (PSAT or PAT) plays a vital role in marine conservation by providing marine biologists with an opportunity to study animals in their natural environments. These are used to track movements of (usually large, migratory) marine animals. A PSAT is an archival tag (or data logger) that is equipped with a means to transmit the collected data via satellite. Though the data are physically stored on the tag, its major advantage is that it does not have to be physically retrieved like an archival tag for the data to be available, making it a viable independent tool for animal behavior studies. These tags have been used to track movements of ocean sunfish, marlin, blue sharks, bluefin tuna, swordfish and sea turtles. Location, depth, temperature, and body movement data are used to answer questions about migratory patterns, seasonal feeding movements, daily habits, and survival after catch and release.

Turtle excluder devices (TEDs) remove a major threat to turtles in their marine environment. Many sea turtles are accidentally captured, injured or killed by fishing. In response to this threat the National Oceanic and Atmospheric Administration (NOAA) worked with the shrimp trawling industry to create the TEDs. By working with the industry they insured the commercial viability of the devices. A TED is a series of bars that is placed at the top or bottom of a trawl net, fitting the bars into the "neck" of the shrimp trawl and acting as a filter to ensure that only small animals may pass through. The shrimp will be caught but larger animals such as marine turtles that become caught by the trawler will be rejected by the filter function of the bars.

Similarly, halfway technologies work to increase the population of marine organisms. However, they do so without behavioral changes, and address the symptoms but not the cause of the declines. Examples of halfway technologies include hatcheries and fish ladders.

LAWS AND TREATIES

International laws and treaties related to marine conservation include the 1966 Convention on Fishing and Conservation of Living Resources of the High Seas. United States laws related to marine conservation include the 1972 Marine Mammal Protection Act, as well as the 1972 Marine Protection, Research and Sanctuaries Act, which established the National Marine Sanctuaries program.

In 2010, the Scottish Parliament enacted new legislation for the protection of marine life with the Marine (Scotland) Act 2010. Its provisions include marine planning, marine licensing, marine conservation, seal conservation, and enforcement.

Organizations and Education

The shore of the Pacific Ocean in San Francisco, California

There are marine conservation organizations throughout the world that focus on funding conservation efforts, educating the public and stakeholders, and lobbying for conservation law and policy. Examples of these include Oceana, the Marine Conservation Institute (United States), Blue Frontier Campaign (United States), Sea Shepherd Conservation Society (international), Frontier (the Society for Environmental Exploration) (United Kingdom), Marine Conservation Society (United Kingdom), Community Centred Conservation (C3), the Reef-World Foundation (United Kingdom), Reef Watch (India), and Australian Marine Conservation Society. Zoox (United Kingdom) is an example of an organization that provides both marine conservation training and professional career development to volunteers who are also working on marine conservation projects such as Green Fins.

On a regional level, PERSGA, the Regional Organization for the Conservation of the Environment of the Red Sea and the Gulf of Aden, is a regional entity which serves as the secretariat for the Jeddah Convention-1982, one of the first regional marine agreements. PERSGA member states are Djibouti, Egypt, Jordan, Saudi Arabia, Somalia, Sudan and Yemen.

EXTINCT AND ENDANGERED SPECIES

Marine Mammals

Baleen whales were predominantly hunted from 1600 through the mid-1900s, and were nearing extinction when a global ban on commercial whaling was put into effect in 1896 by the IWC (International Whaling Convention). The Atlantic gray whale, last sighted in 1740, is now extinct due to European and Native American Whaling. Since the 1960s the global population of monk seals has been rapidly declining. The Hawaiian and Mediterranean monk seals are considered to be one of the most endangered marine mammals on the planet, according to the NOAA. The last sighting of the Caribbean monk seal was in 1952, and it has now been confirmed extinct by the NOAA. The vaquita porpoise, discovered in 1958, has become the most endangered marine species. Over half the population has disappeared since 2012, leaving 100 left in 2014. The vaquita frequently drowns in fishing nets, which are used illegally in marine protected areas off the Gulf of Mexico.

Sea Turtles

In 2004, the Marine Turtle Specialist Group (MTSG), from the International Union for Conservation of Nature (IUCN), ran an assessment which determined that green turtles were globally endangered. Population decline in ocean basins is indicated through data collected by the MTSG that analyzes abundance and historical information on the species. This data examined the global population of green turtles at 32 nesting sites, and determined that over the last 100–150 years there has been a 48–65 percent decrease in the number of mature nesting females. The Kemp's ridley sea turtle population fell in 1947 when 33,000 nests, which accounted for 80 percent of the population, were collected and sold by villagers in Racho Nuevo, Mexico. In the early 1960s only 5,000 individuals were left, and between 1978 and 1991, 200 Kemp›s ridley turtles nested annually. In 2015, the World Wildlife Fund and *National Geographic Magazine* named the Kemp's ridley the most endangered sea turtle in the world, with 1000 females nesting annually.

Fish

In 2014, the IUCN moved the Pacific bluefin tuna from "least concerned" to "vulnerable" on a scale that represents level of extinction risk. The Pacific bluefin tuna is targeted by the fishing industry mainly for its use in sushi. A stock assessment released in 2013 by the International Scientific Committee for Tuna and Tuna-Like Species in the North Pacific Ocean (ISC) shows that the Pacific bluefin tuna population dropped by 96 percent in the Pacific Ocean. According to the ISC assessment, 90 percent of the Pacific bluefin tuna caught are juveniles that have not reproduced.

Between 2011 and 2014, the European eel, Japanese eel, and American eel were put on the IUCN red list of endangered species. In 2015, the Environmental Agency concluded that the number of European eels has declined by 95 percent since 1990. An Environmental Agency officer, Andy Don, who has been researching eels for the past 20 years, said, "There is no doubt that there is a crisis. People have been reporting catching a kilo of glass eels this year when they would expect to catch 40 kilos. We have got to do something."

Marine Plants

Johnson's seagrass, a food source for the endangered green sea turtle, is the scarcest species in its genus. It reproduces asexually, which limits its ability to populate and colonize habitats. Data on this species is limited, but it is known that since the 1970s there has been a 50 percent decrease in abundance.

HISTORY OF MARINE RESOURCES CONSERVATION

Modern marine resources conservation first became globally recognized in the 1970s after World War II in an era known as the "marine revolution". The United States federal legislation showed its support of marine resources conservation by institutionalizing protected areas and creating marine estuaries. In the mid-1970s the United States formed the International Union for Conservation of Nature (IUCN). Through this program, nations could communicate and make agreements about marine resources conservation. After the formation of the IUCN, new independent organizations known as non-governmental organizations started to appear. These organizations were self-governed and had individual goals for marine conservation. At the end of the 1970s, undersea explorations equipped with new technology such as computers were undertaken. During these explorations, fundamental principles of change were discovered in relation to marine ecosystems. Through this discovery, the interdependent nature of the ocean was revealed. This led to a change in the approach of marine conservation efforts, and a new emphasis was put on restoring systems within the environment, along with protecting biodiversity.

OVERABUNDANCE

Overabundance occurs when the population of a certain species cannot be controlled. The domination of one species can create an imbalance in an ecosystem, which can lead to the demise of other species and of the habitat. Overabundance occurs predominately in invasive species. Cargo ships introduce new species into different environments through releasing ballast water into an ecosystem. A tank of ballast water is estimated to contain around 3,000 non-native species.

The San Francisco Bay is one of the places in the world that is the most impacted by foreign and invasive species. According to the Baykeeper organization, 97 percent of the organisms in the San Francisco Bay have been compromised by the 240 invasive species that have been brought into the ecosystem. Invasive species in the bay such as the Asian clam (Corbicula fluminea) have changed the food web of the ecosystem by depleting populations of native species such as plankton. The Asian clam clogs pipes and obstructs the flow of water in electrical generating facilities. Their presence in the San Francisco Bay has cost the United States an estimated one billion dollars in damages.

REFERENCES

1. Ray, G. Carleton (2004) "Issues and Mechanisms", part 1 in Coastal-marine Conservation: Science and Policy. Malden, MA: Blackwell Pub. ISBN 978-0-632-05537-1.
2. "IMPORTANCE OF CORAL REEFS – Coral Reefs – Ocean World". tamu.edu.
3. Trist, Carolyn. "Recreating Ocean Space: Recreational Consumption and Representation of the Caribbean Marine." Professional Geographer. 51.3 (1999). Print.
4. Ngoc, Ninh Thi; Huong, Pham Thi Mai; Thanh, Nguyen Van; Cuong, Nguyen Xuan; Nam, Nguyen Hoai; Thung, Do Cong; Kiem, Phan Van; Minh, Chau Van (2016-09-01). "Steroid Constituents from the Soft Coral Sinularia nanolobata". Chemical & Pharmaceutical Bulletin. 64 (9): 1417–1419. doi:10.1248/cpb.c16-00385. PMID 27321426.
5. Yin, Chen-Ting; Wen, Zhi-Hong; Lan, Yu-Hsuan; Chang, Yu-Chia; Wu, Yang-Chang; Sung, Ping-Jyun (2015-01-01). "New Anti-inflammatory Norcembranoids from the Soft Coral Sinularia numerosa". Chemical & Pharmaceutical Bulletin. 63 (9): 752–756. doi:10.1248/cpb.c15-00414. PMID 26329871.
6. Burke, Lauretta, Liz Selig, and Mark Spalding (2001). "Reefs At Risk in Southeast Asia." World Resources Institute, p. 72.
7. Pandolfi, J. M.; Bradbury, R. H.; Sala, E; Hughes, T. P.; Bjorndal, K. A.; Cooke, R. G.; McArdle, D; McClenachan, L; Newman, M. J.; Paredes, G; Warner, R. R.; Jackson, J. B. (2003). "Global trajectories of the long-term decline of coral reef ecosystems". Science. 301 (5635): 955–8. doi:10.1126/science.1085706. PMID 12920296.
8. "Coral reef destruction and conservation". tamu.edu (18 May 2011).
9. Rodrigo, Raul (1998). Resource at Risk: Philippine Coral Reefs.
10. Key Findings. World Resources Institute Southeast Asia
11. "FIS – Worldnews – Seafood producers tackle ocean acidification at annual Summit". fis.com.
12. Borneman, Eric. "Is It Time to Sell Your SUV? A Tale of Coral Reef Decline". reefkeeping.com.
13. Calcification. reefresilience.org
14. Jenkins, Lekeliad (2010). "Profile and influence of the successful fisher-Inventor of marine conservation technology". Conservation and Society. 8: 44. doi:10.4103/0972-4923.62677.
15. Thys, Tierney (30 November 2003). "Tracking Ocean Sunfish, Mola mola with Pop-Up Satellite Archival Tags in California Waters". OceanSunfish.org. Retrieved 14 June 2007.
16. Block, Barbara A.; Dewar, Heidi; Farwell, Charles; Prince, Eric D. (1998-08-04). "A new satellite technology for tracking the movements of Atlantic bluefin tuna". Proceedings of the National Academy of Sciences of the United States

of America. 95 (16): 9384–9389. doi:10.1073/pnas.95.16.9384. PMC 21347 . PMID 9689089.

17. Hoolihan, John P. (2004-11-16). "Horizontal and vertical movements of sailfish (Istiophorus platypterus) in the Arabian Gulf, determined by ultrasonic and pop-up satellite tagging". Marine Biology. 146 (5): 1015–1029. doi:10.1007/s00227-004-1488-2.
18. Stokesbury, Michael J. W.; Harvey-Clark, Chris; Gallant, Jeffrey; Block, Barbara A.; Myers, Ransom A. (2005-07-21). "Movement and environmental preferences of Greenland sharks (Somniosus microcephalus) electronically tagged in the St. Lawrence Estuary, Canada". Marine Biology. 148 (1): 159–165. doi:10.1007/s00227-005-0061-y.
19. "Turtle Excluder Devices". noaa.gov.
20. Turtle Excluder Device on YouTube
21. Meffe, Gary K. (1992). "Techno-Arrogance and Halfway Technologies: Salmon Hatcheries on the Pacific Coast of North America" (PDF). Conservation Biology. 6 (3): 350. doi:10.1046/j.1523-1739.1992.06030350.x.
22. "Ending Commercial Whaling." ifaw.org (19 February 2015).
23. Carleton, Ray G.; McCormick, Jerry (1 April 2009). Coastal-Marine Conservation: Science and Policy. John Wiley & Sons. ISBN 978-1-4443-1124-2.
24. Noakes, Scott (19 February 2015). "Atlantic Gray Whale: Research: Science: Gray's Reef National Marine Sanctuary". graysreef.noaa.gov
25. "Hawaiian Monk Seal (Neomonachus Schauinslandi)". NOAA Fisheries (15 January 2016).
26. "Caribbean Monk Seal Gone Extinct From Human Causes, NOAA Confirms". ScienceDaily (9 June 2008).
27. Bodeo-Lomicky, Aidan (4 February 2015). The Vaquita: The Biology of an Endangered Porpoise. Createspace Independent Pub. ISBN 978-1-5077-5577-8.
28. "Vaquita". World Wildlife Fund (12 February 2015).
29. "Green Turtle (Chelonia mydas)" NOAA Fisheries. 19 February 2015.
30. "Kemp's Ridley Turtle (Lepidochelys Kempii)." NOAA Fisheries (19 February 2015).
31. "Sushi Edging Pacific Bluefin Tuna Toward Extinction" DNews 19 February 2015.
32. Satran, Joe (19 February 2015). "Pacific Bluefin Tuna Overfishing Has Led To 96 Percent Population Reduction, Study Says" The Huffington Post
33. "European Eel Is Critically Endangered". National Biodiversity Data Centre (19 February 2015).
34. Morris, Steve (30 April 2009). "Eels in Crisis after 95% Decline in Last 25 Years". The Guardian.
35. "Johnson's Seagrass (Halophila johnsonii) – Office of Protected Resources". NOAA Fisheries (1 March 2013).

36. Zacharias, Mark (2014). "Marine Environmental Protection Policy: International Efforts to Address Threats to Marine Biodiversity" in Marine Policy: An Introduction to Governance and International Law of the Oceans. Routledge. ISBN 1136212477.
37. Hinrichsen, Don (2011). The Atlas of Coasts & Oceans: Ecosystems, Threatened Resources, Marine Conservation. University of Chicago.
38. "IUCN, the International Union for Conservation of Nature." IUCN. 12 February 2015.
39. Smith, David. "Ballast Water" MITSG CCR: Marine Bioinvasions (1 January 2006).
40. Choksi, Sejal. "The Hostile Takeover of San Francisco Bay" 1 May 2009. 19 February 2015.
41. Martin, Glen (5 February 2006). "The Great Invaders / A New Ecosystem Is Evolving in San Francisco Bay. We Have No Idea What It Is, or Where It's Going". SFGate.
42. "Foreign Species Invade San Francisco Bay". NPR (11 May 2011).
43. Foster, A.M., P. Fuller, A. Benson, S. Constant, D. Raikow, J. Larson, and A. Fusaro. (2014). Corbicula fluminea USGS Nonindigenous Aquatic Species Database, Gainesville, FL.

Index